Beneath our feet

The caves and limestone scenery of the north of Ireland

BENEATH OUR FEET

The caves and limestone scenery of the north of Ireland

Design and layout by
Corporate Document Services
The Studios
89 Holywood Road
BELFAST
BT4 3BD

Published 2001

This book should be cited as –

Fogg, P. and Fogg, T.
Beneath our feet
The caves and limestone scenery of the north of Ireland
Environment and Heritage Service (Department of the Environment)
BELFAST, 2001

ISBN 1-84123-3471

The Environment and Heritage Service
Project Officer for this publication was
Ian J. Enlander

**ENVIRONMENT
AND HERITAGE
SERVICE**

Environment and Heritage Service
Natural Heritage
Commonwealth House
35 Castle Street
Belfast
BT1 1GU

The Caves and Limestone Scenery of the North of Ireland

BENEATH OUR FEET

THE CAVES AND LIMESTONE SCENERY OF THE NORTH OF IRELAND

CONTENTS

FOREWORD	7
PREFACE AND ACKNOWLEDGEMENTS	9
A CAUTIONARY NOTE & CAVE CONSERVATION CODE	13
CHAPTER 1 LIMESTONE, A ROCK LIKE NO OTHER	15
CHAPTER 2 THE MAKING AND SHAPING OF LIMESTONE	21
CHAPTER 3 THE SPECIAL LANDSCAPE OF LIMESTONE	35
CHAPTER 4 THE EVOLUTION OF THE LIMESTONE OF THE NORTH OF IRELAND	51
CHAPTER 5 OUR LIMESTONE SCENERY ABOVE AND BELOW	59
CHAPTER 6 LIFE ABOVE GROUND AND LIFE IN DARKNESS	91
CHAPTER 7 PEOPLE AND LIMESTONE	107
CHAPTER 8 CAVE EXPLORATION - THE SPORTING SCIENCE	117
CHAPTER 9 KARST AND CONSERVATION	127
WHERE TO EXPLORE LIMESTONE	142
SOME FURTHER READING	146
GLOSSARY	148

The Caves and Limestone Scenery of the North of Ireland

FOREWORD

Limestone is a rock of paradoxes. It is made from the remains of living organisms with a tiny life span, yet it has survived for many millions of years. As a rock, it epitomises hardness, yet it can be dissolved and fashioned – albeit slowly – by mere water.

Its caves evoke particular fascination. On the one hand they can be dark, damp and dangerous, while on the other they are a unique environment for very specialised forms of life. At one time, they were a sanctuary for humans from the worst the elements could offer. The dynamics of the environment are quite different in caves: the cycling of the seasons and night and day may be barely perceptible, but irregular events such as flash floods can have a dramatic influence. Caves are exciting and mysterious. They are also vulnerable.

The cavelands of Fermanagh are unique in Northern Ireland. Owing their origin to a combination of a particular rock and water, they contain a great variety of landforms typically associated with limestone. Sinking streams, caves and pavement characterise large tracts of the area. The landscape is home to a great wealth of important habitat and unusual species, both on the surface and underground. Humans, past and present, have also left their mark, not least in the intriguing variety of names which have been given to caves and potholes, showing the importance of this landscape to people.

There are also other areas with caves: in County Antrim, there are caves together with rivers that disappear only to re-emerge, seemingly from rock, while the environs of Armagh City host small fragments of what may once have been a large cave network. Contiguous areas of RoI share many of the features of Fermanagh's cavelands.

This limestone scenery is important in many ways: to the naturalist, to the caver, to the visitor and not least to the people who live there. It will take the co-operation of all these interested groups to ensure that this fragile environment has a long-term future. One of the requirements for appropriate decision making is information. It is hoped that this book will lead to a better understanding of these areas and help inform all people with an interest in managing this remarkable part of our natural heritage.

Dr. John Faulkner
Director of Natural Heritage
Environment and Heritage Service
Department of the Environment

The Caves and Limestone Scenery of the North of Ireland

PREFACE

This publication has arisen from the Earth Science Conservation Review of Caves and Karst carried out by the Environment and Heritage Service, Department of Environment, in 1995. The review identified a need to inform the public about our karst landscapes which are a delicate part of our national heritage, with importance not only locally but nationally and internationally.

Limestone is a magical rock, producing a special and fascinating karst landscape. Around the world, from countries as diverse as the Mediterranean island of Majorca to the rolling hills of Kentucky, the unifying factor is limestone. Here, in Ireland, we are fortunate to have more than our fair share of this rock. In the north of Ireland, there is plenty to excite our interest and curiosity.

The book is an opportunity to investigate and enjoy the limestone on our own doorstep, not in a distant country. It will look specifically at our landscape, with its own particular character. Within a relatively limited area lies a great variety of classic karst features.

One of the most intriguing things about limestone is its extra, hidden dimension: a subterranean world, a world which evokes in most people a sense of fear and fascination. For scientists, it arouses curiosity - for them, it is a hidden storehouse of information crying out for study. For adventurers, it satisfies the urge for physical challenge and exploration. Everyone has heard of caves and most people know about stalactites and stalagmites, even if they can't remember which one hangs down and which one grows up! But there is so much more to the underground world. Caves are 'windows' which allow us to sneak a glimpse at this hidden environment with its strange creatures and its exquisitely sculpted, water-worn passages. They give us an opportunity to see how water carves a journey through the rock, inextricably linking the limestone surface landscape with the realm beneath.

The intention has been to produce a readable, non-technical book with wide appeal and to make the subject accessible and intriguing. It is aimed at anyone with an inquisitive mind: the people whose homes are on the limestone, the farmer who has its stewardship, the student who examines its finer details and the visitor who appreciates it and moves on with memories. It is hoped that readers will find something to rouse their imagination. The first three chapters give the background to karst and limestone, how it is formed and the features it creates. This sets the scene for an in-depth look at our own limestone. The diagrams help to clarify the complex processes at work in moulding the landscape, while the underground photography illuminates a hidden and often hauntingly beautiful world. Information boxes expand particular themes, and throughout the text specialised terms are highlighted so that the reader can refer to the glossary, which aims to demystify the language of karst.

The object is to bring to the attention of all the importance and vulnerability of our caves and limestone scenery and the need for its appreciation and conservation.

ACKNOWLEDGEMENTS

We wish to acknowledge the assistance of the many people who have helped in the production of this book, particularly Ian Enlander of the Environment and Heritage Service, DOE who has been the driving force behind it.
Also, thanks to Roy Anderson, Les Brown, Michael Coulter, Phil Chapman, Philip Doughty, Brin Edwards, Tony Fogg, Claire Foley, John Kelly, Amanda Lacy, Ben Lyon, Hugh McCann, Mike Simms, Bernard Smith and Tony Waltham.

We are particularly grateful to Joe Walls whose splendid illustrations appear throughout the text, and to Gaby Burns for his detailed maps of Fermanagh. The maps and cave surveys are based upon 'Caves of Fermanagh and Cavan', by Jones, Burns, Fogg and Kelly, and the geology is by Kelly after Hubbard, Brunsom and Mason, (1979).

All other images and illustrations are the property of the authors or Environment and Heritage Service unless otherwise stated. All photographs are by the authors unless credited.

The scanning electron microscopic image of the coccolith was provided by Jeremy Young and Markus Geisen, Natural History Museum, London

Discoverer Series map and aerial photograph is reproduced by permission of the Ordnance Survey of Northern Ireland on behalf of the Controller of Her Majesty's Stationary Office © Crown copyright 1989 and 1994. Permit number 1643.

The permission of the Geological Survey of Ireland the use the solid geology map of Ireland is gratefully acknowledged.

A final word of thanks to the landowners of the limestone and caves for their ready welcome, hospitality and interest.

Pamela and Tim Fogg,
May 2001.

The Caves and Limestone Scenery of the North of Ireland

A CAUTIONARY NOTE

CAVES ARE HOSTILE PLACES FOR HUMANS. EXPLORATION OF THEM SHOULD ONLY BE UNDERTAKEN BY THOSE WHO ARE PROPERLY EXPERIENCED AND EQUIPPED.

FOR THE FOOLHARDY, ILL INFORMED AND POORLY EQUIPPED, CAVES PRESENT HAZARDS WHICH ARE POTENTIALLY LIFE THREATENING.

SHOW CAVES, SUCH AS THE MARBLE ARCH CAVES IN FERMANAGH, OFFER THE CASUAL VISITOR A SAFE MEANS OF VIEWING THIS UNDERGROUND WORLD.

■ *A CAVE CONSERVATION CODE*

Our caves are fragile places. Their features take hundreds and more often, many thousands of years to form. Their inhabitants, the bats and tiny insects, are vulnerable and easily disturbed.

Irreparable damage can be done, often unwittingly, by people who enter without first learning about caves and their conservation. A stalactite, once broken, may never re-grow; it has no beauty or worth outside a cave. Cave sediments are storehouses of information which, if disturbed, are no longer of value. Cave life once upset, may never recover. With one careless movement, something unique can be lost forever.

A cave is a special environment and a wonderful place for scientific research and personal adventure. There is a huge responsibility on cave visitors to learn about and respect the world in which they are privileged but temporary voyeurs.

■ *ACCESS*

THE DESCRIPTION OF ANY SITE IN THIS BOOK DOES NOT INDICATE THAT A RIGHT OF WAY OR RIGHT OF ACCESS EXISTS. EVERY SITE IS OWNED BY SOMEBODY AND PERMISSION TO ENTER THE LAND SHOULD BE SOUGHT FROM THE LANDOWNER.

The Caves and Limestone Scenery of the North of Ireland

Beneath our Feet

Limestone, a rock like no other

1

The Caves and Limestone Scenery of the North of Ireland

LIMESTONE, A ROCK LIKE NO OTHER

Limestone is the underlying skeleton of a dramatic and unique landscape. Deep gorges, rocky hillsides, disappearing rivers and mysterious caves are just some of the fascinating features which let us know we are in a **karst** area.

Karst is the general term used to describe scenery composed of soluble rock with underground drainage flowing through cave passages. As limestone is the most widespread soluble rock, covering one twelfth of the world's land surface, nearly all karst landscapes are formed on limestone.

The word 'karst' comes from the Slovene word *kras* which means dry, bare, stony ground. It was in the area between Slovenia and Italy that geologists studied and named some of the typical limestone features. Now these Slovenian words are used world-wide to describe features seen in limestone landscapes from the hills of Fermanagh to the rain forests of Borneo.

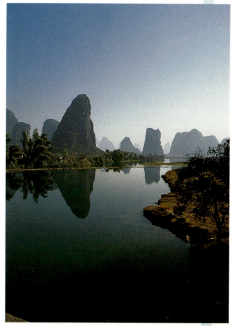

Fig. 1 Not an imaginary scene from a Chinese painting. The extraordinary landscape of classic tower karst in Guilin, China.

■ WHAT IS A CAVE?

A cave is defined as any naturally formed hole in the ground big enough for humans to enter. Here, in the north of Ireland, when we think of caves we probably think of sea caves, carved by the pounding action of waves. Or perhaps Belfast's Cavehill comes to mind. The caves that give this landmark its name are formed in volcanic rock.
Caves are also found where great slabs of rock have slumped or collapsed such as Cove Cave in the Mourne Mountains.
In other parts of the world, caves also occur in glaciers, where they constantly change as the ice moves and melts.
However, most sea, lava or glacier caves pale into insignificance when we consider caves formed by the dissolving away of limestone. It is in this rock that most caves of the world are found. Other soluble rocks like gypsum do develop caves but it is in limestone that we find most of the record-breakers.

The Caves and Limestone Scenery of the North of Ireland

Fig. 2 The dramatically desolate landscape of the Burren in County Clare contrasts with the lakes and limestone hills on the Sligo/Leitrim border.

Throughout the world, there are extensive areas of karst. China, for example, has more limestone than the rest of the world put together, while France has such a thickness of limestone that it has some of the deepest known caves in the world. However, Ireland also has its share. A glance at a geological map of Ireland (Fig. 3) gladdens the heart of any karst enthusiast, as it shows that limestone underlies almost half of the island. This represents one of the largest single areas of limestone in Europe. Much of it is, however, hidden from view by a blanket of overlying sediments.

Not only are there huge areas of karst scattered around the world, there are also variations in its appearance world-wide, depending on a combination of factors: different types of limestone, past and present climates, recent glaciations and so on. Even within this small island of Ireland, the karst landscape of one county may look different from the karst of another.

The Burren area of County Clare is famous for its wild and spectacular karst scenery, scoured by glaciers. The bare rock landscape is almost lunar but it is far from barren. Its cracks and hollows nurture an amazing variety of plant life including many rare and unusual species.

Further north, there is an extensive arc of limestone extending through the counties of Sligo, Cavan, Leitrim and Fermanagh. The limestone of Antrim, Tyrone and Armagh is limited but there is just enough visible to let us glimpse some karst features. County Antrim's small but intriguing outcrop of limestone is coded green rather than blue on the geological map as it is of a different geological age and has been created and shaped by a different series of geological events.

The Caves and Limestone Scenery of the North of Ireland

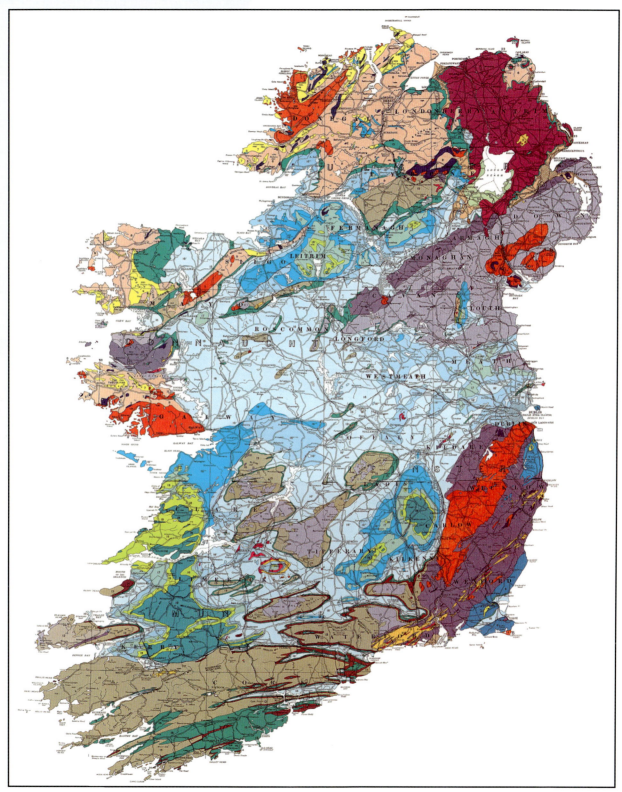

Fig. 3 Limestone (colour-coded blue) underlies almost half the island of Ireland.

The Caves and Limestone Scenery of the North of Ireland

It is from the two counties of Fermanagh and Antrim that much of the material in the following chapters is drawn. Fermanagh will feature in particular, as it has the largest area of visible limestone. We will look at how limestone is made and what gives it its special qualities. We will look at how it has evolved, to give us those sinking rivers, bare hillsides devoid of flowing water and mysterious cave entrances.

This limestone world is not, of course, just one of rock and water; it is a landscape which shelters life. Plants and animals have shaped and adapted their existence over time to survive and flourish where soils are thin and water scarce. Human beings too have developed ways of co-existing with this challenging environment; their presence on the limestone can be traced back for thousands of years. Stone Age burial sites and early Christian hilltop settlements merge with the present day farms and villages.

We can experience the special nature of karst scenery by simply walking or driving through it and looking around. If we want to delve further, we can visit a show cave, where the underground world has been made safe by ingenious engineering.

However, for a full explanation of this special landscape, we must draw on knowledge from many fields of science including geography, geology, hydrology, archaeology and biology. We must also go to a breed of scientist cum adventurer willing to penetrate past the twilight of the cave entrance and venture underground, the **speleologist**.

The Caves and Limestone Scenery of the North of Ireland

Beneath our Feet

The making and shaping of limestone

2

THE MAKING AND SHAPING OF LIMESTONE

To unravel the intricacies of the surface and underground scenery of our own limestone areas, we must first think about its bones - the limestone itself, how it was made and what it was made from. We must investigate the changes in the environment that have moulded it over the millions of years since its creation.

ROCK TYPES OF THE EARTH

Three main rock types make up the earth's crust. They are classified by the way they have been formed. Igneous rocks are created by the cooling and solidification of hot liquid rock (magma), either on the earth's surface or deep down. Sedimentary rocks are formed by the settling and consolidation of material deposited by rivers, wind or the sea. Metamorphic rocks come about when igneous or sedimentary rocks are changed in character and appearance by heat or pressure.

Limestone is a sedimentary rock. It usually originates through the accumulation of the remains of animals, plants or microbes in a marine environment. It is defined as containing at least fifty per cent **calcium carbonate** ($CaCO_3$), a common mineral extracted from sea-water by organisms and included in their bodies as skeletons or shells.

The process of limestone formation is a slow one; generation after generation of marine organisms die and settle on sea-beds. Layer upon layer of limey deposits build up - perhaps a small amount in one year but over hundreds of millennia, great depths of sediment are formed. Within the accumulating layers, chemical deposition also takes place. For example, the pure mineral form of limestone, **calcite**, may crystallise out from calcium carbonate-rich waters. With pressure from the sediments above, coupled with the cementing effect of crystallisation, the deposits are consolidated into **beds** of limestone rock.

Some remains of organisms encased in the sediments remain intact through the whole process of change from soft muds to rock and can be seen clearly as **fossils** in the rocks today.

HOW FOSSILS FORM

The term fossil (from the Latin *fossum*, to dig) at one time meant anything dug up from the earth. It is now applied only to the remains, imprint or trace of an animal, plant or microbe. The tracks and burrows of worms, the faint marks made by the fronds of waving seaweed or the bones of a dinosaur are all found as fossils. Under the microscope we can see, often perfectly preserved, the fossil remains of micro-organisms.

The Caves and Limestone Scenery of the North of Ireland

Fig. 4 A fossilised fan of coral (Siphonodendron) preserved for millions of years and now exposed in the wall of a cave on East Cuilcagh. For an illustration of the same fossil look at Fig. 56.

Fig. 5 It is strange to see part of an ancient beach so perfectly preserved hundreds of metres above sea level. Although technically not fossils, these ripple marks are perfectly preserved in layers of Fermanagh sandstone.

■ **WHAT ARE THE CONDITIONS FOR FOSSILISATION?**

Though many rocks teem with fossils, these represent just a small proportion of the earth's inhabitants through the millions of years of geological time.

Only those creatures whose bodies include hard resistant material are likely to be fossilised. Their soft parts are eaten, or else decay very quickly. It is rare for a complete body to be preserved.

Particular environmental conditions are also necessary for fossilisation. Most commonly, dead organisms are incorporated in the undisturbed sediment layers of quiet lagoons, sea basins or the still waters of swamps.

Less common fossils range from tiny flies preserved in amber (a mineral formed from the resin of trees), to huge mammoths entombed in glacier ice and ancient humans preserved in peat bogs.

Casts and moulds of organisms are also common forms of fossils in which the shell or bone no longer exists but whose shape can still be seen in the hardened sediment.

Complete replacement of animal or plant material by a mineral (e.g. calcite or silica) also occurs in some cases of fossilisation. This results in petrified remains (from the Latin *petra* meaning rock and the Greek *facere* to make like). Petrified forests, for example, are some of the most beautiful and detailed of these types of fossils.

Some of the fossil corals found in Fermanagh are **silica** replacement fossils. They can now be seen sticking out of the main limestone blocks as the silica is more resistant to weathering than the limestone.

As no two places on sea floors experience the same conditions, it follows that there is a great variation in the composition of the sediments being laid down at any one time. For example, in one area, huge coral reefs may flourish and deposit coral debris, so forming limestone crammed with fossils. In another, silica-rich sponges may dominate resulting in a limestone containing silica nodules called **chert** or **flint**. Near a land mass, rivers bring silt and sand into the sea to settle and mix with the calcium carbonate-rich sediments, so making sandy limestones. In the clear, still water of a lagoon, pure, uniform limestone accumulates. The variations in types of limestone are endless.

As figure 6 oppsosite shows, changes in the environment over long stretches of geological time also influence the composition of the sediments. As a result, beds of different character lie on top of each other. The breaks between the beds mark distinct changes and are called **bedding planes**.

When first formed, beds and bedding planes lie parallel to the original sea floor. Massive forces within the earth can work to tilt and fold the beds into mountain ranges, exposing the rocks to weathering. Uplifting of the land, coupled with the removal of the weight of

The Caves and Limestone Scenery of the North of Ireland

Fig. 6 The influence of changing environments on the formation of Carboniferous limestone.

KEY
A. Ammonoid
B. Brachiopod
C. Crinoid
D. Bryozoa
E. Coral
F. Gastropod
G. Coral
H. Swamp forests

350 MILLION YEARS AGO

THE CARBONIFEROUS PERIOD

300 MILLION YEARS AGO

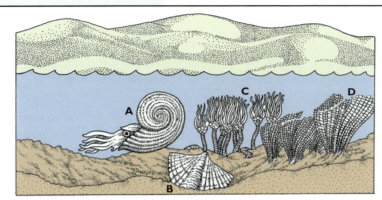

1. The climate is hot and dry. The clear, warm waters of a lagoon swarm with life. Layers of lime-rich sediments accumulate on the sea floor.

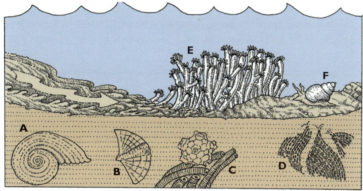

2. Conditions change. Storms bring flows of muddy sediment across the sea bed burying the remains of earlier life. Colonies of coral and gastropods are more suited to these murky waters.

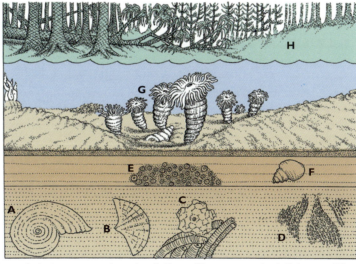

3. More change. On land, great swampy forests flourish. These will become seams of coal. Under water, different creatures colonise the brackish waters. With the weight of overlying layers, the sediments are slowly turned into solid limestone.

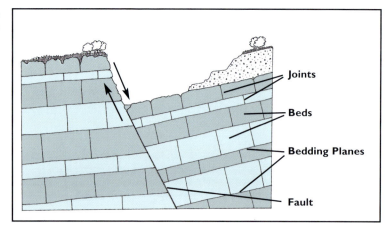

Fig. 7 Beds of limestone fractured by joints and faults.

the upper layers by erosion, releases pressure. This causes fracturing of the beds so adding vertical fissures or **joints** to the pattern of horizontal bedding planes. Continued earth movement over time can also shear the beds and displace great blocks of rock, one against another. The massive fractures which result are called **faults**.

Volcanic action may further complicate the story, either by injecting molten rock from deep in the earth's crust into the rock or by erupting and pouring out over its surface.

And always the great erosional forces of ice and water are working to re-shape the surface layers, altering and sculpting the limestone landscape.

All these major changes, over an immense time span, are recorded in the layers of sedimentary rock and its surface scenery so that now, when we glance at a limestone hillside or quarry face, we have a snapshot of a piece of the earth's history before our eyes. Every bed, every joint, every variation in the shape, texture and colour is evidence of life, change and upheaval.

■ LIMESTONE'S SPECIAL PROPERTIES

Now that we know how limestone is made and what it is made of, we can investigate why a landscape composed of limestone has such an intriguing peculiarity : why water disappears from its surface and travels underground, creating cave passages as it goes. This vanishing act can be explained by the fact that the rock itself has some special properties.

At first glance, these properties seem almost contradictory because on the one hand, limestone is a strong rock, yet it also has weaknesses: it is soluble and it has fractures. You only have to look at its prominent cliffs or acknowledge the demand for its quarried stone to realise that limestone must be a mechanically strong rock. In fact, it has been calculated that a pillar of limestone could be built to height of 4,000 metres (more than 2 miles) high before starting to crush under its own weight. If a piece is examined under a microscope the key to its strength can be seen. It is made up of tightly interlocking crystals. This structure came about when limestone was being laid down as sediments on the sea floor. The dissolving of calcium carbonate from the remains of sea life and its crystallisation within the sediments was a crucial part of the strengthening process.

It is difficult to imagine that rock, seemingly as resilient as limestone, can actually be soluble. Its vulnerability lies in the fact that it is composed of calcium carbonate, which is easily dissolved by rain and soil water, both of which are slightly acidic. The tight structure of the rock means that it is impervious; in other words, it does not act like a sponge, water cannot just sink into it anywhere. However the beds and fractures throughout the rock mass act as weaknesses, which collect and channel the acidic water, so concentrating it and enhancing the dissolving action.

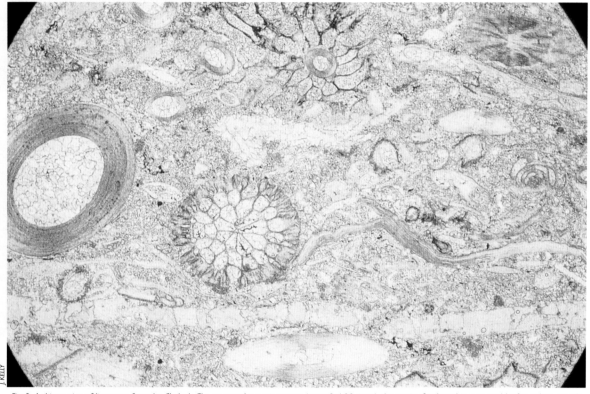

Fig. 8 A thin section of limestone from the Cladagh Glen seen under a microscope (magnified 20 times). It contains fossils and is cemented by fine calcite crystals.

The Caves and Limestone Scenery of the North of Ireland

Much of the water flowing on to limestone in the north of Ireland comes from peat bogs. This water is particularly acidic because bogs are dominated by bog mosses (Sphagnum) which have particular powers of water acidification. The mosses swap hydrogen ions for minerals necessary for their growth and the released hydrogen ions acidify the water, making it more **aggressive**, and capable of dissolving greater quantities of limestone.

To give some idea of the weathering effect of aggressive water, it has been calculated that in Fermanagh, bare limestone is lowered by approximately 50 mm (2 inches) in 1,000 years and in a tropical karst region, by as much as a staggering 1,100 mm (44 inches) in the same length of time.

The wearing away of the rock by solution isn't the only process of erosion going on. Mechanical weathering , known as **corrasion**, also plays a part in the shaping and sculpting of limestone and is simply the wear and tear inflicted by rock debris as it is carried along by flowing water.

■ DISSOLUTION AND DEPOSITION

Limestone is carved by water but pure water alone cannot dissolve enough of limestones main component, calcium carbonate, to create a karst landscape.

The essence of chemical erosion (**corrosion**) of limestone is a chemical reaction that occurs in the air and in the soil; carbon dioxide gas is absorbed by water from the air and from the decomposition of material in the soil to form dilute carbonic acid. This acidic water is known as **aggressive water** and it attacks the calcium carbonate in the limestone, dissolving and removing it, and producing the soluble mineral, calcium bicarbonate.

When water laden with dissolved limestone (calcium bicarbonate) leaves the confines of a fissure and emerges into an open cave passage, the chemical reaction is reversed. Much of the carbon dioxide in the water bubbles away into the cave air, forcing calcium carbonate to precipitate gradually from the solution to be deposited as the pure mineral form of calcium carbonate, **calcite**. The most familiar of these calcite deposits are stalactites and stalagmites but the variety is endless. We will look in more detail at the array of cave decorations in the next chapter.

As with many natural chemical reactions, the dissolution of limestone and deposition of calcite is often far more complex than these simple reactions imply.

The Caves and Limestone Scenery of the North of Ireland

■ **THE CHEMICAL PROCESSES SHAPING LIMESTONE (Fig. 9).**

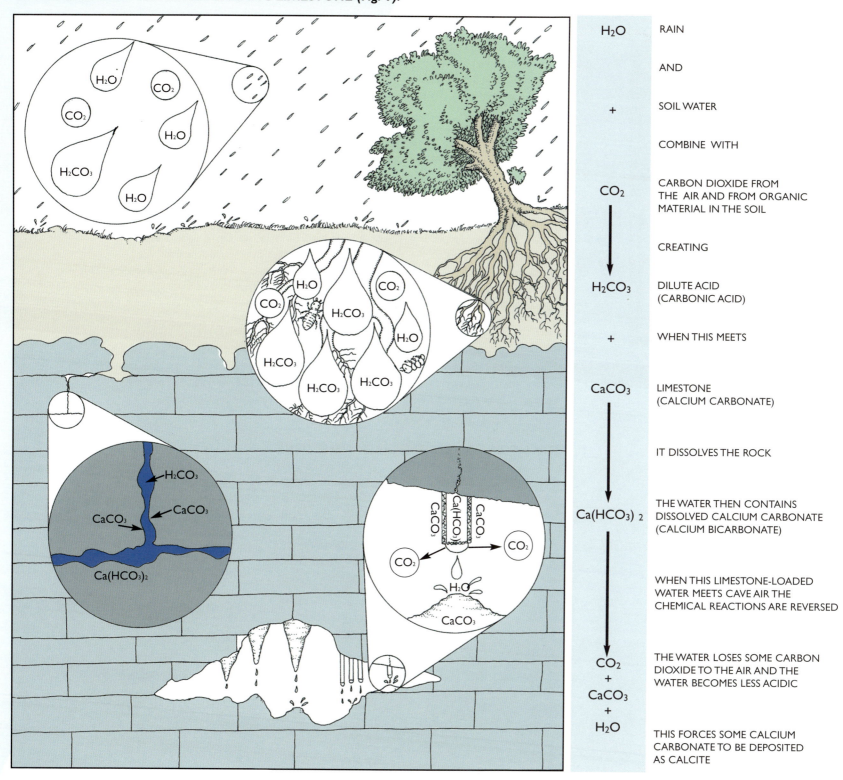

The Caves and Limestone Scenery of the North of Ireland

In the safety of a laboratory, it is easy to test the vulnerability of limestone to the action of acidic water. When a concentrated acid is dribbled on to a piece of limestone, the rock will bubble and fizz as carbon dioxide is produced and the surface is etched and dissolved before our eyes. This experiment confirms the sample as a **carbonate rock**.

These chemical reactions can be seen in action in many other ways in our everyday lives. People whose water supply comes from limestone areas may wonder why their soap does not lather, or why a scaly white deposit builds up in their kettle. This water is known as hard water as it has high levels of calcium carbonate in solution in the form of calcium bicarbonate.

Small stalactites may often be noticed growing under bridges and from old stone work. These occur because calcium-rich water is leaching from the mortar. This is a slightly different chemical reaction to the one that forms stalactites in caves but it achieves a similar end result. The chemical composition of mortar means that the process happens faster than in caves and so these 'man-made' stalactites may grow in a comparatively short space of time.

In our modern world, cars, factories and oil and coal fired power stations are pumping out all manner of waste gases, particularly sulphur dioxide and oxides of nitrogen, which, when released into the atmosphere, produce acid rain. To witness the far-reaching effect of chemical weathering, we only have to look up at an old limestone building. Surfaces are pitted and damaged, while statues are often altered beyond recognition.

Fig. 10 A limestone statue slowly being eaten away by acid rain.

■ HOW WATER TRAVELS THROUGH LIMESTONE

Rain falls on limestone, rivers flow onto limestone and then disappear from view. Water runs off the blocks of limestone, etching the surface as it goes and is funnelled and concentrated into its cracks and fissures.

As the diagrams in Figure 11 show, water finds the easiest route down through the system of fissures, following and enlarging the main beds, joints or faults. The geology of the rock dictates which routes the water takes.

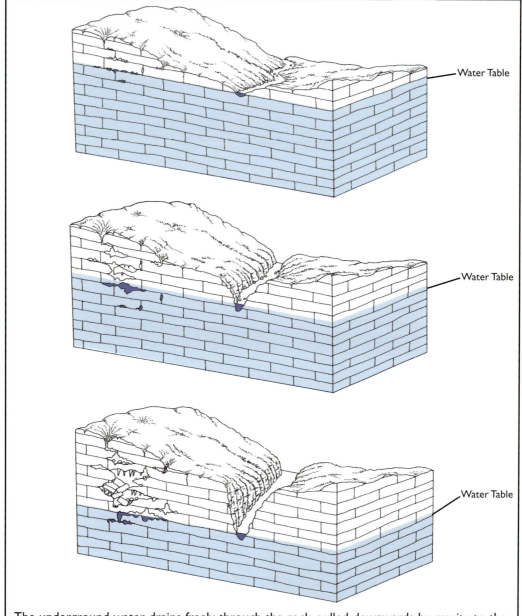

Fig. 11 The valley deepens and the water tables drops.

The underground water drains freely through the rock, pulled downwards by gravity to the water table, which, as the name suggests, is the top of the body of groundwater completely filling every space within the rock mass. It reflects the general level of water in the area, and varies with the amount of water feeding down to it.

Over the great expanses of geological time the water table level changes. As rivers, or more dramatically, glaciers, erode valleys, the floors are deepened and the water-table drops. Underground, upper passages are abandoned as cave rivers seek new routes at lower levels.

The Caves and Limestone Scenery of the North of Ireland

■ KARST HYDROLOGY

Caves provide the only way to study 'face to face' the underground flow of water through a limestone mass. However, the vast part of the drainage system in a limestone area is inaccessible to humans. Studying how and where it flows underground is known as karst **hydrology**.

The tracing of underground water and linking particular river sinks with their spring risings initially came about because the people living and working on the limestone needed to know more about their water supply. They wanted to predict how reliable a source it was, and where it came from.

There are a number of ways to trace water's subterranean journey: chemicals such as common salt or the spores of mosses can be put in where the water sinks and recorded at which spring or springs they reappear on the surface. In the past, water was traced, often unintentionally, when chaff from grain was washed into rivers. The chaff would later be noticed emerging at a particular spring. Nowadays, harmless dyes such as fluroscein are generally used and dye detectors, usually small bags of charcoal, are put at all possible spring risings. If the water containing the diluted and now invisible dye flows over the charcoal, it can be detected using laboratory equipment and so a positive link between a particular river sink and its rising can be made.

Other studies can be carried out on these subterranean drains: the speed of flow can be measured, the temperature recorded, and the water chemistry analysed. All this information helps reveal more about hidden supplies of water.

Fig. 12 Harmless fluorescent dye is poured into a river as it disappears underground.

THE KARST WATER CYCLE (Fig. 13).

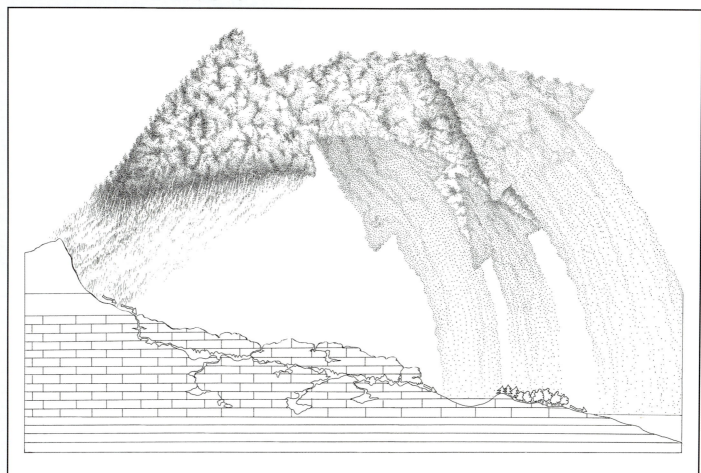

The movement of water through the karst system is part of the global circulation of water from the oceans to the atmosphere, thence to the land and so back to the oceans. Evaporation takes place from the oceans, lakes and rivers, while plants breathe out, or transpire, sending more water into the atmosphere for re-circulation as rain or snow, so continuing the water cycle. In areas of karst, an extra dimension is added to the normal water cycle. Water continues to flow through the cycle but for a period, while it is in caves and in the rock as ground water, it is not lost back into the atmosphere. Karst areas are therefore major storage areas within the water cycle.

The Caves and Limestone Scenery of the North of Ireland

Beneath our Feet

The Caves and Limestone Scenery of the North of Ireland

The special landscape of limestone

The Caves and Limestone Scenery of the North of Ireland

THE SPECIAL LANDSCAPE OF LIMESTONE

There are numerous clues which help karst detectives deduce that they are in a limestone landscape: bare white rock, river valleys without rivers, gaping pot holes - the detective just has to know the signs. A dry, grassy hollow may correspond 30 metres below to a subterranean river cascading along a cave passage; many of the surface features will confirm that water is flowing under the ground rather than over it. Similarly, underground, a multitude of features relate both to the evolution of the cave and to the surface landscape.

The individual features do not exist in isolation; they are all part of an interacting, interconnecting system which is still changing and evolving at this moment.

ABOVE GROUND

The surface landscape in a limestone region is made up of an intriguing array of physical features, some dramatic, some subtle.

■ SINKHOLES

When a surface stream meets limestone, it is not long before it finds a chink in the rock armour, a geological weakness such as a crack or fissure, which it can exploit with its chemical and mechanical erosive powers to create a **sinkhole** or swallow hole.

Fig. 14 A river plunges down a pothole on the slopes of East Cuilcagh.

The shape and size of the sinkhole depends on the major weaknesses in the rock. Spectacular sinks are created when the river encounters a vertical joint or fault; a pothole will develop, gulping the water down shafts, which, in Ireland, can be as deep as 100 metres. Variations on the theme are numerous but it is usually possible to see the geological feature which has controlled the way in which the water is channelled into the underworld. Many sinkholes lie in the base of enclosed hollows called dolines, described below.

■ DOLINES

Depressions, called **dolines**, are the land forms which let you know, more than any other, that you are in a limestone region. If there is a doline above ground, then there must be something happening beneath.

They look like craters and have no outlet for surface downhill drainage. Instead, they act like enormous coffee filters, percolating water into the underground drainage system. Their size varies from a few metres to hundreds of metres in diameter.

Fig. 15 Dolines are key features of a karst landscape. Rather than flowing downhill on the surface, water drains into a central sinkhole in circular depressions.

The Caves and Limestone Scenery of the North of Ireland

As Figure 16 shows, they are formed in a number of different ways: by solution and widening of fissures and bedding planes; by collapse, usually of the roof of the underlying cave passage; by subsidence of the overlying material into the joints, sometimes revealing the bedrock beneath; or by a combination of all these. The name doline comes from the Slovene meaning 'closed valley' but they are often known by the descriptive term **shakehole**, particularly if the bedrock is exposed. In Fermanagh, they are sometimes called cup holes and many of them incorporate the Irish word *lag* meaning sag, hollow or cavity e.g. Legacapple, *lag an chapaill*, hollow of the horse.

Fig 16 Dolines are most commonly formed by:

SOLUTION, when joints are progressively widened by solution,

COLLAPSE, when the roof of the cave passage beneath falls in,

SUBSIDENCE, when overlying sediment, such as glacial deposits covering limestone, are constantly subsiding and feeding down the widened joints.

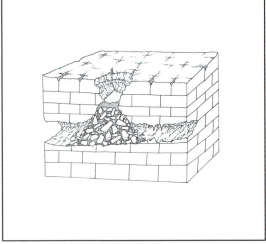
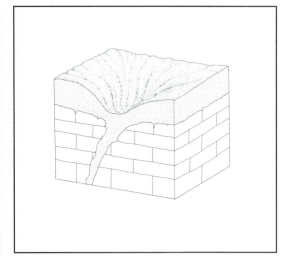

The Caves and Limestone Scenery of the North of Ireland

LIMESTONE PAVEMENT

Areas of **limestone pavement** stand out white and stark in the landscape and as the name suggests, they are bare rock surfaces. They are one of the legacies of the stripping and scouring action of ice. (For a photograph of classic limestone pavement look at Figure 3). Features which owe their origin and character to the effect of the ice on limestone, are known collectively as **glaciokarst**.

In the thousands of year since the ice finally retreated, most other glaciated landscapes have, by now, generated a new covering of soil. But this is often not the case with limestone; the constant dissolving and removal of the rock means that soils do not develop easily and so the surface may remain dramatically exposed.

Limestone pavement is, however, relatively limited in extent in the north of Ireland compared to other karst areas, as the ice, having scraped the rock clean, then re-covered it with a layer of clay.

As soon as the rock was laid bare by the ice, rainwater set about carving the surface, like a sculptor transforming a block of stone, and now these natural rockeries are fretted with a variety of rainwater-etched features. These features measure from a few millimetres to a few metres and are collectively called by the German term **karren**. **Clints** and **grykes** are the most obvious and probably the best known. Grykes are the water-widened fissures which separate blocks of limestone called clints. Sliced into the clints are channels or runnels which act as mini water-slides 'chuting' the surface water underground. As Figure 18 shows some have sharp edges-sometimes razor sharp-if developed on bare rock and others have rounded edges if they have developed beneath a covering of soil.

Fig.17 An area of limestone pavement near Knockmore covered with a thin layer of soil.

Fig. 19 The dissolving action of the acidic water has worked its magic to create a piece of natural rock art. The miniature basins in the surface of the clints are known as solution hollows.

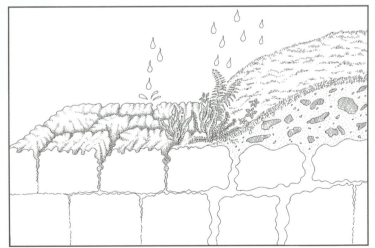

Fig. 18 Rain water is funnelled into the joints in the bare pavement, sharply etching the rock as it goes. Wider cracks are produced when the limestone is under a covering of soil. Seepage water filtering through soil and plant roots becomes highly acidic and can dissolve up to six times more limestone (calcium carbonate) than rain water falling directly on to bare limestone.

■ DRY VALLEYS

Now we come to another group of common, and somewhat puzzling karst features. **Dry valleys** look exactly like their name suggests - valleys with no rivers flowing in them. Obviously, water at some time must have carved them. They have various origins. In some cases, their vegetated valley floors confirm that it is a very long time since they were active water carriers. They may have been carved when the fissures in the limestone were blocked with ice and torrents of glacier melt water flowed on the surface. The valleys later lost their rivers when the land thawed out and the limestone could behave as limestone once more. The rivers sank down into the rock to flow beneath the surface, perhaps only a few metres beneath, perhaps hundreds of metres.

In other dry valleys it is obvious that rivers still occasionally flow. This happens in times of heavy rainfall, when the underground fissures and caves in the area are completely flooded and the water has no choice but to flow on the surface.

Fig. 20 A dry valley. No water flows on the surface. The valley floor is marked by a line of rushes.

On a similar theme, a **blind valley** occurs at a point where the valley literally stops dead where its stream disappears underground, often at the base of a cliff. The stream may sink at different points upstream from its final end point depending on how much rainfall there has been; consequently, sometimes the river bed may be completely dry, at others the river may be flowing on the surface right up to the 'blind' end. If there has been sufficient rainfall or if the sink has become blocked with flood debris, a lake can form at the cliff base.

Gorges are steep-sided and often spectacularly narrow valleys. Some may be attributed to the collapse of a cave roof but most were probably carved by those torrents of glacial melt water cutting rapidly and with tremendous force down through the limestone.

■ SPRINGS

Fig.21 The Aghinrawn River disappears underground at the base of Monastir Cliff.

Springs are the outlet points for the underground drainage systems, and, because cave streams converge underground, there are fewer springs than sinks. They are also called **risings** or **resurgences**. Water emerges from the limestone mass for many different reasons. Perhaps the simplest is when it encounters a layer or bed of impervious, water-resistant rock. Local quirks of geology like weaknesses, folds or faults in the rock may also bring the water to the surface. Similarly, natural damming underground or fluctuations in the local water table may also cause the water to resurge.

Springs can be all sizes, from a trickle to a mighty river. Sometimes, the water wells up from a deep pool, the flow never drying up, even after long summer droughts. In other cases, it gushes out in a cascading waterfall as if it can't wait to get back to sunshine. Some may bubble up from stream or lake-beds or even from the bottom of the sea.

Some caves have a number of risings, depending on how much water is entering and

The Caves and Limestone Scenery of the North of Ireland

flowing through the underground system; if, for example, there is torrential rain, the cave may not be able to accommodate the volume of water and so the river exits at a higher point.

■ VANISHING LAKES

Turlough is the Irish name for a dry lake but they are, in fact, seasonal lakes, occurring in lowland limestone. It is perplexing to come across them in the countryside in summer time, as they have all the look of a lake yet they are often completely empty. If we come back in winter they may be full to the brim. Our knowledge of limestone would lead us to believe that a lake cannot collect in bare, exposed limestone; it would be like trying to collect water in a sieve. However, if the sieve is lying in a basin of water then it will also contain water. This is how turloughs occur. They have no distinct streams flowing in or out of them. Instead, they owe their changing level to fluctuations in the local water table as the ground water level rises and falls depending on how much rainfall there has been in the area. Often in winter, the lakes may freeze over, except for the small areas over the springs which remain unfrozen. Turloughs, as well as being intriguing geological features, often shelter some special flora which we will look at in more detail in Chapter 6.

Fig. 22 A rising at Hanging Rock Nature Reserve undergoes a dramatic transformation. Usually it is bone dry but after periods of particularly heavy rain it starts to flow, spurting water as if a giant tap has suddenly been turned on.

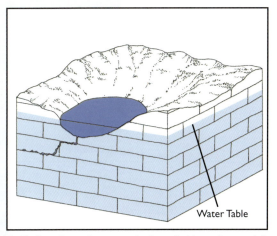 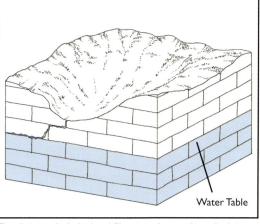 

Fig. 23 Depending on the season turloughs can be lakes or dry, grassy hollows. They drain into the bedrock and fill again as the groundwater level rises during periods of wet weather.

The Caves and Limestone Scenery of the North of Ireland

KARST OF LONG AGO

The conditions for the development of karst also existed in eras long before the one which shaped our present-day landscape. Limestone scenery was formed, with the same typical karst features such as dolines and caves which we see today; then it was covered over and fossilised under younger rocks, perhaps by flows of lava or because of a rise in sea level. Nowadays, it is sometimes possible to see evidence of these ancient limestone features buried by successive layers of younger rocks. Such features are referred to as **palaeokarst** or **fossil karst**.

THE UNDERGROUND ACTION

We can now look at how the surface features in a karst landscape relate to the world underground.

To understand and describe caves and the features within them, we have to look at how they evolve. The process is called **speleogenesis** and as the word genesis implies, it starts at the absolute beginning or inception of underground drainage and continues right through to the cave's eventual destruction.

CAVE EVOLUTION

This process can be seen as having three distinct stages. The first is called the **initiation** stage, when the water is in minuscule cracks. The second is regarded as the main stage, as it covers the enlargement of tiny networks of tubes into passages which are cave-sized-big enough for humans to enter. The third and final stage is the process of degradation which may involve the total in-filling of the cave with sediment, its collapse or even its complete removal by erosion.

Fig. 24 Palaeokarst in the cliffs of County Antrim. A large cone-shaped depression in the upper surface of the limestone was infilled, then covered by the outpouring of lava. The infill is composed of a kind of volcanic ash mixed with flints. The flints may have originally weathered out from the limestone to give this clay-with-flints material.

Fig. 25 A cave evolves. Tiny water filled tubes combine to become cave passages as the water level drops.

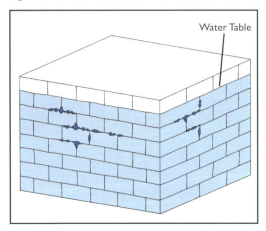
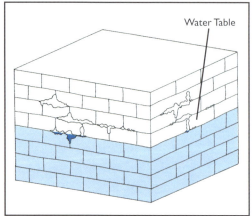
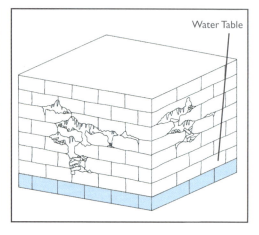

To complicate matters, all three stages might be going on at the same time in different levels of a complex cave system or even within one section of passage. And to confuse matters further, a lucky cave can have its ageing process reversed! A passage, for example, may have been filled with sediment only to find it washed out and reworked by a youthful, new stream.

STAGE 1: THE LIQUID BEGINNINGS

Having worked its way into the spaces in the limestone, what happens to the water next? In these early stages of the development of the underground drainage route, the water spreads out in a thin layer known as **laminar flow** along bedding planes or other fractures in the limestone. Over time, the chemical solution by the aggressive acidic water will widen the cracks allowing increased **turbulent flow**. Differences within the limestone itself, for example, the presence of fossils or impurities, will make areas within the cracks less susceptible to corrosion. Purer areas of limestone will open faster, capturing and increasing the flow in what then become definite tubes. These networks, called **anastomoses**, may contain hundreds of tubes. They are, in effect, micro-cave passages with a typical diameter of around 10cm. When one channel offers easier flow conditions it increases in size at the expense of the others, which are then abandoned and left dry.

Fig. 26 A network of half-tubes. We are able to see this early stage in cave development because the cave ceiling has collapsed.

In addition, when two tubes meet, there is twice as much water to work the magic. Fine grains of sand and other abrasive material are suspended in this greater flow, which acts like liquid sandpaper. They add to the erosive capabilities of the water by mechanically eating away at the limestone, a process called corrasion. The speed of flow in these water-filled passages is slow and the rate of development is correspondingly slow. At this stage all spaces in the rock are completely filled with water, in conditions known as **phreatic**.

Erosion in phreatic passages takes place around the whole outline so that the resulting shape tends to be smooth and more or less round. The fissure, which was the original geological control of the passage, will still dominate the shape, giving in some cases beautiful elliptical cross-sections.

Fig. 27 A phreatic passage. This passage is now dry but its smooth rounded outline lets us know that water once filled it completely and eroded its entire surface.

The Caves and Limestone Scenery of the North of Ireland

Fig. 28 Often cave walls are elegantly patterned with spoon-shaped hollows called scallop marks which have been eroded by the turbulent eddies in flowing water. In water-filled phreatic passages the movement of water is slow and so the scallops marks are large, whereas free-flowing water in vadose passages moves faster, carving smaller hollows.

If the water which made the scallop marks has long abandoned the passage, it is still possible to tell in which direction it flowed, as the upstream end of the scallop marks have steeper, sharper lips. Scallop marks are therefore one of the clues which help build up a picture of how a particular cave was formed.

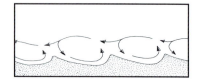

■ HOW LONG DOES A CAVE TAKE TO DEVELOP?

The initiation of caves takes a long time; a million years may pass while water finds a route through limestone. However, as the route opens and the flow pattern of the water changes, the rates of solution and erosion increase and the speed of development picks up. Cave scientists estimate that a passage diameter can increase by one metre in ten thousand years.

STAGE 2: THE CAVE DEVELOPS

Just as streams above ground join to create a river, so the underground drainage tubes collect to create major passages which carry the flow efficiently down through the rock. As the water table drops (Fig.11), the speed of flow through the limestone increases and passages at higher levels in the limestone are no longer completely water-filled. Passages only partly filled with water are known as **vadose**. Below the water table, erosion continues to take place in phreatic passages. These water filled passages are often referred to as **sumps** or siphons.

Within vadose passages, water flows freely down slopes or vertically under gravity; therefore, the passages have air above the water surface. Erosion of the passage walls will only occur at, or below, water level. Distinctive and often dramatic passage shapes develop, such as sweeping canyons and shafts. Spray from waterfalls may etch surrounding walls, adding further water-worn decoration to the naturally sculpted rock.

The cave at this stage is highly active. Water levels react to the weather above ground and so vary according to the amount of rain falling. Just like man-made drains, passages may fill to the roof when a storm is raging and empty again as the storm eases. These episodes of flooding and draining give little opportunity for calcite or other sediments to build up to any great extent. Sediments can only start to accumulate in a particular part of a cave when the water abandons it in its search for new passages in progressively lower levels. When a section of cave is completely abandoned and is inactive as far as the main water flow is concerned, it is called **relict cave**.

Fig. 29 A cave river cascades down a waterfall, a typical feature of a vadose passage.

Fig. 30 When a phreatic passage is no longer completely water-filled, a vadose groove, then a canyon, is carved in its floor making a classic keyhole-shaped cross-section. Flowing water has long since abandoned this passage.

Fig. 31 These photographs show the original geological weaknesses which guided the water through the limestone: bedding planes, and vertical fissures.

Fig. 32 A drop of water drips from the tip of a delicate straw stalactite.

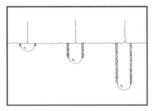

Fig. 33 This stalactite and stalagmite will eventually meet up to form a column.

STAGE 3: DECORATION, DECLINE AND FALL

We now enter the stage for which caves are possibly most well known. It is during this phase that stalactites and stalagmites get a chance to develop. Water, the great maker and shaper of caves, now becomes their interior decorator.

Calcite precipitates out of the calcium carbonate-laden water, as it emerges from the cracks and into the cave. It is deposited anywhere the water is dripping, trickling and dropping: on the walls, roofs and floors. The detailed chemistry of calcite deposition has been explained in Chapter 2 but, simply, when water laden with dissolved limestone emerges from a crack in to the cave passage, some of its carbon dioxide bubbles away. It cannot, therefore, hold as much dissolved limestone and so it leaves some behind. Stalactites and stalagmites may be the best known but there are many other forms of calcite deposition, with an infinite variety of colours, shapes and sizes. Collectively, these chemical sediments are called decorations or **speleothems**. This term is derived from the Greek for cave spelaion and deposit thema.

The simplest form of speleothem is the straw stalactite. These delicate tubes grow from the cave roof. Single drips of water deposit a crystal ring of calcite, always 5mm in diameter, the size of a water droplet. Water continues to feed down the developing hollow tube adding crystals to the end point before splashing to the floor. The straw may develop into the characteristic carrot-shaped stalactite if additional calcite builds up on the outside of the straw cylinder or if a blockage in the tube forces water to seep through its walls.

Water droplets falling from the end of stalactites to the cave floor have already given up some of the dissolved limestone, but on bursting as they hit the floor, more carbon dioxide is lost to the cave air and more calcite is deposited. A stalagmite is born! A calcite column occurs when the stalactite growing down meets the stalagmite growing up.

The Caves and Limestone Scenery of the North of Ireland

■ HOW LONG DO STALACTITES TAKE TO GROW?

Unfortunately, it is almost impossible to answer this question. Each stalactite forms at its own pace. The pace is controlled by the flow rate of the feed water, the amount of calcium bicarbonate in the water, and the air conditions. It is however, possible to find out how old a particular stalactite is by radio isotope dating. This technique involves drilling a small sample from the stalactite. The sample is then analysed to measure the ratio between the isotopes of the element (e.g. uranium or carbon) incorporated in the calcite on its formation. The proportions alter precisely over time and the natural clock is read.

Both straw stalactites and the larger stalactites will eventually, if they continue to grow, become too heavy for themselves and break off. Looking at the cross section of a broken stalactite is like looking at a tree trunk: the growth rings can be seen. However, the stalactite rings reflect changes in the character of the feed water rather than the accurate growth per year, as seen in the tree rings.

The longest known stalactite in the world is 28 metres. It hangs from the roof of a cave in Brazil. In Ireland, the stalactite in Pol-an-Ionain in County Clare is still one of the longest in the world at 6.5 metres. In the north of Ireland, some of the biggest reach just over 2 metres from roof to tip, as in the Marble Arch System.

We may not be able to say exactly how long stalactites take to grow but we do know that the process is a slow one. Touching their delicate growing tips can halt their growth or cause them to break off altogether.

Fig. 34 One of the longest stalactites in the Marble Arch Caves.

The simple gravity led speleothems have weird and wonderful cousins called **helictites** or eccentrics. They come in a bewildering array of shapes and sizes: sometimes they are curly, almost like hairs; in other places they are perfectly horizontal, defying gravity completely. All this eccentricity is the result of deformed crystal growth. The theories as to how they actually grow are almost as varied as the formations themselves. Ideas range from the influence of draughts blowing along passages, to pressure within cracks, to impurities or electrical charges.

Fig. 35 Helictites defy gravity and grow out from the wall.

The Caves and Limestone Scenery of the North of Ireland

Fig. 36 Cave water deposits calcite as it trickles down the sloping ceiling and along the hem of this curtain hanging in Pollnagollum Cave.

Fig. 37 Cave pearls. Taken out of the cave they loose all their lustre and become aesthetically worthless. Their beauty, as with all speleothems, depends on their remaining undamaged in the cave where they belong.

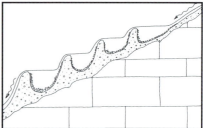

Fig. 38 When water fills any hollows then overflows, the thin film of water overflowing the edge gives off more carbon dioxide and so more calcite is deposited. Over time, these rims may grow into dams known as **gours**, and a terraced effect of crystal-lined **gour pools** may develop.

Fig. 39 **Flowstone** is formed when the water pours across sediment banks or rocks depositing calcite as it goes. This grand flowstone almost blocks the entire floor of Skreen Hill Three in the Marble Arch System.

Rivulets of calcium carbonate-saturated water running down walls and inclined ceilings produce curtain-like draperies of calcite known as **curtains** which are often beautifully banded with different colours.

Occasionally, within pools, **cave pearls** may appear. Just as their name suggests, they look like the pearls of an oyster, round and creamy white and they form in much the same way. Calcite builds up around a grain of sand. The constant agitation of the shallow water, perhaps from water droplets, rolls and shapes the pearl.

The range of speleothems is infinite: from helictites growing on stalagmites to curtains hanging from stalactites. Impurities in the water may add many different colours to the brilliant white of pure calcite. How each forms is the subject of much discussion and reams of scientific papers. We will see more of our own speleothems in the next chapter.

Speleothems, having first added colour, beauty and interest to the cave, may eventually choke it completely. Furthermore, in this final phase, sections of the roof and walls may start to collapse as passages become larger. Essentially, the rock loses support as the passage size increases and eventually it collapses. The way it breaks down will be governed by the fissure systems in the limestone. Some caverns can reach immense proportions before this starts to happen. It all depends on the geology.

In addition to calcite sediments, caves can also be choked with infillings of mud, sand, gravel and boulders, which have entered the cave in flowing water or even been injected by tongues of ice or torrents of melt water during glacial phases. These sediments are more commonly seen in the underground landscape as river banks, or they may make up the floor material of dry passages. The patterns and layers within the banks hold information about the conditions in which they were laid down. Pebbles all lying in the same direction give an indication of the water flow direction. Chaotic jumbles of big and small stones suggest that massive floods carried along these heavy loads, then dropped them as the water subsided. Layers of sorted material tell of fluctuations in type and content of flow.

These materials hold valuable clues about past environments. Cave muds may, for example, hold information about shifts in the earth's magnetic field while pollen grains within the muds can tell us which plants thrived in bygone eras.

Complete degradation of a cave system leaves us with nothing to enter and explore. However, it is not unusual for passages which have been infilled to be excavated and re-activated by a new input of water. It is easy to imagine a situation in which drainage patterns are changed by human action or by natural events. New streams may find their way into previously abandoned or choked caves and the process can begin all over again. Drying and shrinkage of the sediment over long periods of time may also allow a blocked passage to be re-entered.

When we look up at the lofty, vaulted roof of a cave passage, it can be difficult for us to imagine that change is on-going. But the sound of water dripping from the tip of a stalactite or the distant trickle of a cave stream, should be enough to remind us that the cave is still evolving, perhaps growing in one section, while being gently choked with calcite in another.

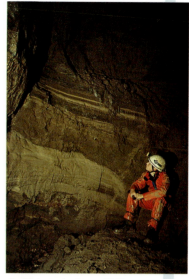

Fig. 40 Layered sediments fill this passage in the Marble Arch System.

Fig. 41 Huge rock slabs have collapsed from the roof of this cave and now litter the floor. Breakdown along lines of weaknesses may continue until the passage is completely blocked.

The Caves and Limestone Scenery of the North of Ireland

Dividing the evolution of caves into three stages from their liquid beginnings through to their decline, is just one way of imposing some order on the complex and hidden processes at work. But one thing is certain, time and lots of it, is the essential ingredient in the process of cave development.

Fig. 42 Water is constantly working to re-shape the limestone landscape. Underground, calcite is deposited anywhere water is dripping, trickling and dropping: on the walls, roofs and floors.

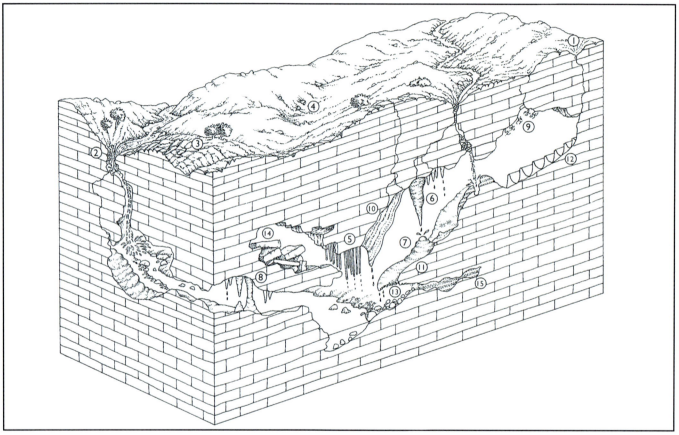

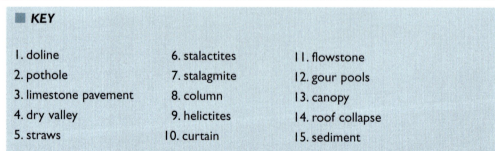

■ **KEY**

1. doline
2. pothole
3. limestone pavement
4. dry valley
5. straws
6. stalactites
7. stalagmite
8. column
9. helictites
10. curtain
11. flowstone
12. gour pools
13. canopy
14. roof collapse
15. sediment

The Caves and Limestone Scenery of the North of Ireland

Beneath our Feet

The evolution of the limestone of the north of Ireland

4

The Caves and Limestone Scenery of the North of Ireland

FROM A TROPICAL SEA TO A ROCKY HILLSIDE

Long time periods are difficult to imagine. The 4,500 million years of the earth's existence as a solid planet almost defies comprehension. During the millions upon millions of passing years, massive changes have occurred on it. The tiny piece of the earth's surface which we know as Ireland has been lifted up and its mountains have been eroded; it has been inundated by seas and has reverted again to land; it has been shaken by volcanic activity and ice has formed in great sheets over it. At the same time, it has been moved around over the surface of the globe.

Around 570 million years ago, Ireland was situated south of the equator, roughly where South East Asia is today. Since then, the drift of the continental plates has brought it around the globe to its present position, 4000 kilometres east and 55 degrees of latitude north of the equator.

350 million years ago (early in the geological period called the Carboniferous), Ireland had completed roughly half its journey. It had moved east and north but still lay in the tropics. This was an arid phase for Ireland but it also experienced occasional dramatic floods which washed sediments into the surrounding shallow warm seas. One such sea occupied the area we know as Fermanagh and Tyrone (Figure 6 gives some idea of what was going on). The sediments building up on this sea floor were a product of these changing environmental conditions - layers of mud sandwiched between layers of marine animal skeletons. The resulting sedimentary rocks are thinly layered limestone full of fossil marine creatures, interspersed with thin layers of siltstone. In Fermanagh today, these are called the Glencar Limestones.

Fig. 43 Coral islands in the South China Sea. How Fermanagh looked approximately 330 million years ago, a tropical paradise!

As millions of years passed, the character of the tropical seas changed. Lagoons with coral reefs dominated. The floods of early times moderated. The sediments laid down in this period gave rise to much purer limestone: Fermanagh's Dartry Limestone. In some places, these thick layers of grey limestone have nodules of brown or black rock called chert, incorporated within them. Chert originates chiefly from organisms which lived in the tropical seas and used silica as their body-building mineral. Unlike calcium carbonate, silica is not soluble in water. As a result, chert lumps can often be seen protruding from eroded limestone surfaces.

The Caves and Limestone Scenery of the North of Ireland

Fig.44 The small rounded knolls rising above the general level of the landscape were once mud banks in tropical lagoons.

In the fertile waters of the lagoons, banks of calcium-rich mud built up. It is not known for sure how or why these mounds accumulated. Some may have built up around algal colonies, while others may have been shaped by currents and as a result they have few, if any, beds. Today, they appear in the landscape as small hills or knolls.

Towards the end of the Carboniferous period (about 290 million years ago), a lifting up of the land-mass resulted in a change from lagoons to sandy estuaries and swamps. Again, these changes in environment resulted in a change in the materials which were being deposited and are evident in the Fermanagh rocks today : layers of shale and sandstone lie on top of the Dartry Limestone. The most striking are the Glenade Sandstones which stand out as cliffs on the highest points in the Fermanagh countryside.

As you will have realised, the rocks are given names, such as Dartry and Glencar, to describe specific types of sandstone or limestone with particular characteristics. They are often named after the places they were studied, so, for example, Dartry Limestone was first examined in the Dartry Mountains in County Leitrim. It is still called Dartry Limestone when found in Fermanagh or elsewhere, if it has the same characteristics and is part of the same geological sequence.

Fig. 45 Three different limestone types found in the north of Ireland.

Glencar: thin beds of shales and limestones.

Ulster White Limestone: pure white with nodules of insoluble flint protruding from it.

Dartry: pure, pale grey with thick beds.

The Caves and Limestone Scenery of the North of Ireland

Fig. 46 *The geological time scale together with simplified stratigraphical columns for Fermanagh and Antrim.*

PERIOD	AGE (millions of years)	MAIN NORTHERN IRISH ROCK TYPES	GEOLOGICAL ENVIRONMENT	EVOLUTION OF PLANT AND ANIMAL LIFE
QUATERNARY	1.6	Blown sand, peat; Sand, gravel and boulder clay	Beach, lake, river and peat bogs; Glacial deposits from glaciers and ice-caps	Modern humans. Woolly mammoth.
TERTIARY	65	Clay, lignite; Basalt lava	Marsh and lake; Fissure eruptions	Earliest Hominids. Widespread mammals. First primitive apes. Main bird groups present.
CRETACEOUS	135	White Limestone	Marine conditions with accumulating planktonic material	Extinction of dinosaurs, plesiosaurs and ammonites. Early flowering plants.
JURASSIC	205	Mudstone with minor limestone	Marine conditions	Dinosaurs dominant on land, plesiosaurs in oceans and pterosaurs in air. Early mammals.
TRIASSIC	250	Red mudstone with salt; Sandstone	Coastal lagoons. Shallow water in continental desert	First dinosaurs and large marine reptiles. First flies. Ammonites common.
PERMIAN	290	Limestone and sandstone	Desert conditions with occasional marine influence	Reptiles spread on land. Insects spread. Conifers common.
CARBONIFEROUS	355	Sandstone and minor coal; Limestone	Coastal lagoons; Marine with coral reefs	Amphibians spread. Shark-like fish. Early trees and reptiles appear.
DEVONIAN	410	Sandstone, mudstone Conglomerate	Continental desert with periodic floods	First amphibians. Fern-like plants on land.
SILURIAN	438	Shale and sandstone	Mainly deep marine conditions	First land plants. Armoured, jawless fish common.
ORDOVICIAN	510		Occasional volcanic activity	Early fish appear. Graptolites, trilobites and brachiopods common in oceans.
CAMBRIAN	570	Not present		Dominance of trilobites in seas and development of early shelled forms.
PRECAMBRIAN	4,600	Metamorphic rocks: schist	Sedimentary and some igneous rocks changed by later metamorphism	Early multi-celled animals. Early bacteria and algae.

AGE	SUMMARY OF CO. ANTRIM
Tertiary	Basalt
Cretaceous	Chalk
Jurassic	Clay

ROLE OF ROCK IN KARST SYSTEM
- Protects the soluble rock
- Hosts the karst features
- Generally acts as the impermeable base to the karst zone

SUMMARY OF CO. FERMANAGH	AGE
Sandstone and shale	Carboniferous
Dartry Limestone	Carboniferous
Glencar Limestone	Carboniferous

The Caves and Limestone Scenery of the North of Ireland

COUNTY ANTRIM GETS ITS LIMESTONE

150 million years after the Carboniferous period, environmental conditions were perfect for the formation of another, different limestone. Although probably occurring over a wide area, this limestone is now mainly seen in Co Antrim. During the Cretaceous period, the earth's plate of which Ireland was a part, had drifted further north and lay where Spain is now. The surrounding seas were cooler than the tropical seas of the Carboniferous period and were home to masses of marine plankton. Plankton are small, almost microscopic organisms which, in common with the tropical corals, build their skeletons from calcium carbonate extracted from sea water. Myriads of tiny single-celled organisms called coccolithophores produce surface scales of calcium carbonate, coccoliths, which often make up the bulk of the sedimentary material. The resulting sedimentary rock formed from the plankton debris is called chalk. It contains flint similar to the chert found in the limestones of Fermanagh.

It would appear that the Antrim chalk underwent another change which makes it different from other chalk. A process of recrystallisation made it harder and less open in texture, giving it those special properties so important to the development of karst and caves as described in Chapter 2. By comparison, many chalks elsewhere are structurally weak and porous, and so rarely develop cave passages. The Antrim chalk is referred to as Ulster White Limestone.

■ HOT ROCK

The next significant geological event in the Antrim area occurred some 60 million years ago during the Tertiary period, when volcanic activity began. Fracturing of the earth's crust, associated with the opening of the North Atlantic, resulted in lava pouring over the Ulster White Limestone in Antrim.

Fig. 47 The intricate beauty of coccoliths can only be appreciated when they are seen under a microscope.

Fig. 48 A basalt lava flow in Hawaii today. County Antrim was experiencing similar volcanic action 60 million years ago.

Fig. 49 A 'liquorice allsort' effect was created by the addition of this black layer of basalt on top of the dazzling white rock in the cliffs around the north coast of Antrim.

The Caves and Limestone Scenery of the North of Ireland

Volcanic action also affected Fermanagh. Molten rock did not flow out and engulf the surface of the land as it did in Antrim but instead lava was forced through faults in the limestones creating basalt dykes.

We will see later that these walls of insoluble rock play an important part in influencing the way water flows through the landscape.

■ THE ICE AGE: SCULPTOR OF OUR PRESENT DAY LIMESTONE LANDSCAPES

In comparison to the rest of the geological time scale, the next great landscape moulding event in Ireland, the Ice Age, was recent and relatively short, but it was extremely significant. From approximately two million years ago to as recently as 10,000 years ago, this area was subjected to great swings in climate: periods of intense cold, interspersed with intervals of warmer weather. There were at least three major cold periods through which the ice waxed and then waned. In fact, although this period is called the Ice Age, Ireland was more often free of ice than covered with it. During the warmer phases, great rivers of melt water poured in torrents from the thawing and retreating mass of ice.

Fig. 50 The volcanic dyke which runs like a wall through the lower slopes of Cuilcagh Mountain.

The last major ice sheets developed only 25,000 years ago with an enormous thickness of ice centred in the Lough Neagh basin. Glaciers from Scotland moved down into Antrim, while the great ice sheet covering Donegal extended into Tyrone and Fermanagh. Hundreds of metres in depth, these masses of ice engulfed the landscape; in some places they completely obscured it, while in other places, just the peaks of the highest mountains were left to protrude through their icy blanket.

Before the ice came, much of Fermanagh's limestone lay protected from erosion under the sandstone and shale beds. The crushing weight of millions of tonnes of moving ice removed parts of these protective layers and exposed the limestone. Laid bare to the elements the limestone did not, however, react like a karst rock when temperatures rose and melt waters came. The fissures which are so important in guiding water into the limestone were still blocked with ice or glacial debris, inhibiting the development of underground drainage. Erosion was concentrated above ground, allowing the development of normal surface drainage patterns. This helps explain why, in our present day limestone landscape, we see enigmatic features like dry valleys (Fig. 20).

Fig. 51 During the Ice Age this part of Ireland must have looked like Antarctica does today.

Any cave systems that were already in existence before the onset of the Ice Age would probably have been held in a state of suspended animation during the cold phases. Then, during the warmer interglacial times, the powerful physical force of huge volumes of melt water would have reawakened them, accelerating the process of their development, and also initiating new cave systems.

The Caves and Limestone Scenery of the North of Ireland

Fig. 52 As the ice moved from east to west it deepened the Lough Macnean valley and steepened its sides.

During the cold phases, the cave resurgences in the valleys were blocked by ice and the water table was consequently temporarily raised; but, as the glaciers melted, the water table dropped to lower levels, causing cave rivers to cut faster down through the limestone, leaving dry passages above (Fig.11).

The power of the ice was immense; spurs of land blocking its path were simply 'bulldozed' aside. All the rock and debris that was scooped up and pushed along by the ice, had, eventually, to be dropped. The clay and gravels which today cover much of Ireland's limestone were deposited by the ice.

The tremendous power of the ice can be appreciated further when we see massive boulders called erratics which were transported away from their point of origin and set down on different bedrock. It is not hard to understand why folk legend attributes their mysterious presence in the landscape to the power of giants.

When the entrances of caves finally became free of ice blockages, sediment-laden melt waters from the glaciers thundered through the underground passages leaving huge sandstone boulders and banks of muds, sands and gravels (Fig. 75).

Fig. 53 A sandstone erratic from the slopes of Cuilcagh Mountain lies on the Marlbank limestone and protects the limestone beneath it from weathering. The height of this plinth gives some idea of how much the limestone surface has been lowered since the last ice melted.

The shape of our landscape and the complexities of our caves owe much to the effect of the Ice Age. Of course, it is not only the great forces of nature which have given us the countryside we know today. In the few thousands of years since the retreat of the last ice sheet to the present day, it is man who has affected and shaped the scenery more than any other single element. From the quarrying of whole hillsides to the clearance of the great forests and burning of fossil fuels, man's period of influence may have been short but it has been profound; it can take only days to alter what nature has worked for millions of years to put together.

Beneath our Feet

The Caves and Limestone Scenery of the North of Ireland

Our limestone scenery above and below

5

OUR LIMESTONE SCENERY ABOVE AND BELOW

The scene has been set; now we can take a detailed look at our own limestone scenery.

Fermanagh has the greatest concentration and variety of limestone features in the north of Ireland. The Antrim karst is unique but limited; however, having said that, it is possible that only a fraction of the cave passages in its Ulster White Limestone have yet been explored. Tyrone and Armagh also have a hint of karst which, although only just visible, still merits a mention.

COUNTY FERMANAGH

The Fermanagh karst is divided, only politically, from a larger band of limestone which forms an arc around the central upland mass of north Leitrim and Cavan. As the map shows, the karst areas extend around this roughly circular upland, from Sligo's Benbulben Mountain in the north west, through north Leitrim and Cavan, to the south-western part of Fermanagh. The exposed limestone continues south and west, just south of Lough Allen to the Bricklieve Mountains in south Sligo.

Fig. 54 Fermanagh's limestone is part of a larger band of limestone extending into the surrounding counties.

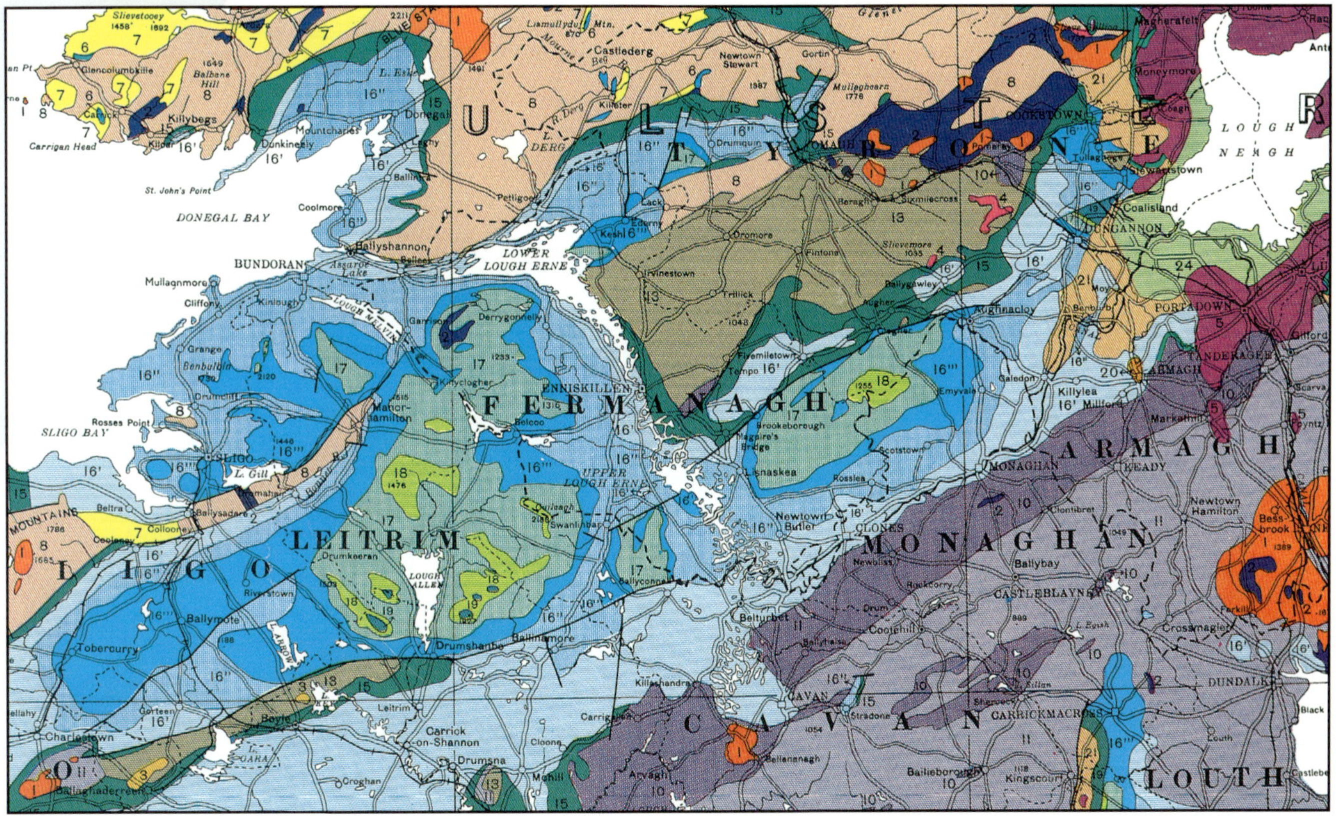

The Caves and Limestone Scenery of the North of Ireland

Fig. 55 *An artist's eye-view of Fermanagh, looking south west.*

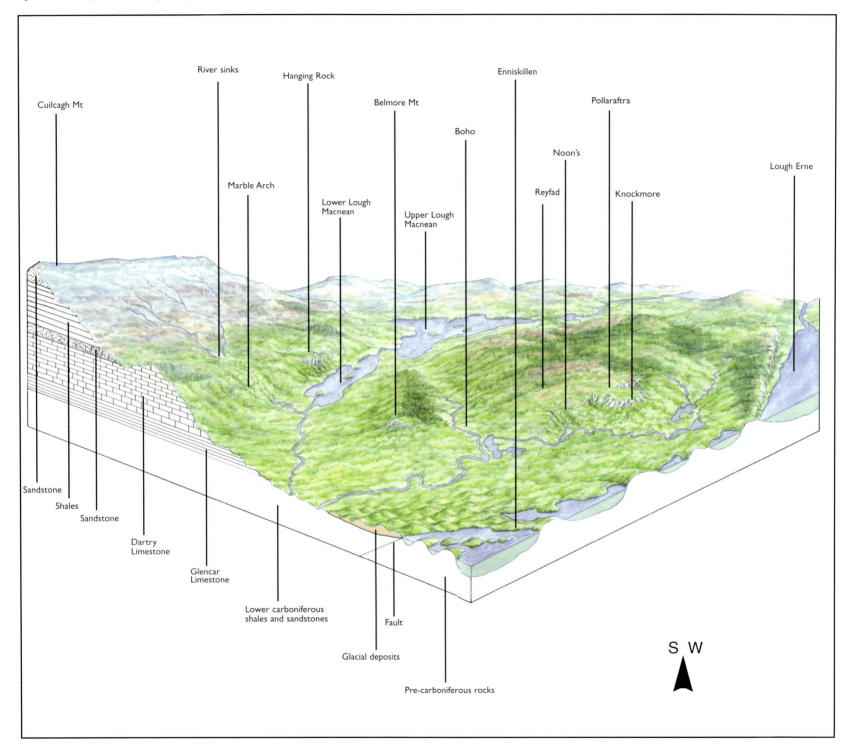

The Caves and Limestone Scenery of the North of Ireland

In the Fermanagh area, there are about 52 square kilometres of karst landscape, virtually all developed on the Dartry Limestone. There are two distinct belts of visible upland limestone bisected by the Lough Macnean Valley, one south of the valley, north of the Cuilcagh Mountain ridge, and the other lying north of the Macnean Loughs, the uplands of Ballintempo. The geology of these two upland masses is basically the same, as they are split only by the relatively new, glacier-carved valley.

Lying outside these two main belts are a number of other karst sites in Fermanagh which are also worth describing.

The youngest rocks in the Cuilcagh/Ballintempo region, the sandstones, are at the top of the local geological pile and cap the tops of the highest areas, while the oldest rocks, including the Glencar Limestones, lie exposed in the valley floors.

The summit ridges of Cuilcagh and the Ballintempo uplands are sandstone. Under the sandstone are thin layers of shale, more sandstone, and mudstone. These beds lie over the Dartry Limestone which in turn sits on top of the Glencar Limestone. (Chapters 2 and 4 describe the creation of these limestones in more detail.) Faulting and volcanic activity has slightly confused this simple story in some places, as we will see shortly when we describe specific sites. As a result of this layered geological sequence, the surface landscape, or topography, which we see today is one of flat topped, sandstone-capped plateaux stepping down to the lake-studded lowlands.

The main karst areas lie on and within the level terraces of Dartry Limestone which divide upland from lowland. Streams, flowing from the plateaux, sink into the Dartry Limestone and then re-surface at the base of the steep slopes in the valley floors. These springs tend to be along the junction between the Dartry Limestone and the lower, older Glencar Limestone as the thin beds of water-resistant shale within the Glencar Limestone cause the water to emerge from underground. Classic surface karst features, such as dolines and dry valleys, described in the previous chapter, are seen in abundance on these terraces and slopes.

Many different fossils can be found in these layers of sedimentary rock, particularly in the limestones. The common types are shown on the following page.

The Caves and Limestone Scenery of the North of Ireland

Fig. 56 *Siphonodendron* viewed from the side and viewed from above. The same fossil can look very different depending on how it is found in the rock. (For a photograph of this fossil see Fig.4).

■ THE LIMESTONE FOSSILS OF FERMANAGH

The most common fossils in the Dartry and Glencar Limestones are bryozoa, corals and brachiopods. It is also possible to see gastropods and crinoids.

The bryozoan, *Fenestella* is the commonest fossil in the limestones. Known as sea mat, it is the classic net-like fossil in the mudbank limestones. Tiny organisms build lace-like colonies from calcium carbonate. Similar creatures can be found living on the stones, shells and seaweed of our beaches today.

The corals are of two main types, solitary and colonial. The solitary corals look a bit like cones, horns or cylinders if their whole body can be seen, whereas end on, they appear as circles or oblongs with hub and spoke-like patterns within.
The colonial corals are spectacular. They grow from a central stem and branch out like a tightly packed small bush. If looked at from above, they appear as numerous circles with jagged inner edges. The one in the diagrams is called *Siphonodendron*.

The brachiopod fossils are often found as perfect shells, commonly called lamp shells as they resemble the shape of an Aladdin's lamp. They consist of two unequal shells; the larger of the two has a small hole in it which was occupied by the creature's fleshy stalk or arm. The word brachiopod means arm-foot.

The snail-like shells found fossilised in the limestones are gastropods - single, spiral shells similar to whelks and winkles. Fossil crinoids (from the Greek for lily) or sea lilies are, in fact, the remains of marine animals. It is common to find the stalks of the crinoids which look like jointed tubes, while the softer head and foot parts are less commonly preserved.

Fossils can also be found in the shale and a few plant fossils can be seen in the sandstone.

Fig. 57 Fossil crinoid stalks.

■ THE CUILCAGH KARST

The three kilometre ridge of Cuilcagh Mountain with its imposing line of cliffs along its northern slopes, is a distinctive landmark. Its summit is the highest point in the region at 668 metres above sea level and can be seen from as far away as County Limerick, 140 kilometres to the south.

From the mountain's flat top to its foot at Lough Macnean is a sweeping panorama of geological history. All the landscape's past environments are captured in the rocks: swamps and deltas in the sandstone at the top; lagoons with fields of sponges, corals and the weird mud banks in the Dartry Limestone sandwiched in the middle; and beneath, the estuaries and mud flats in the Glencar Limestone.

Fig. 58 Looking south across Lower Lough Macnean to the northern slopes of Cuilcagh Mountain.

The Caves and Limestone Scenery of the North of Ireland

Fig. 59 The Cuilcagh Mountain region.

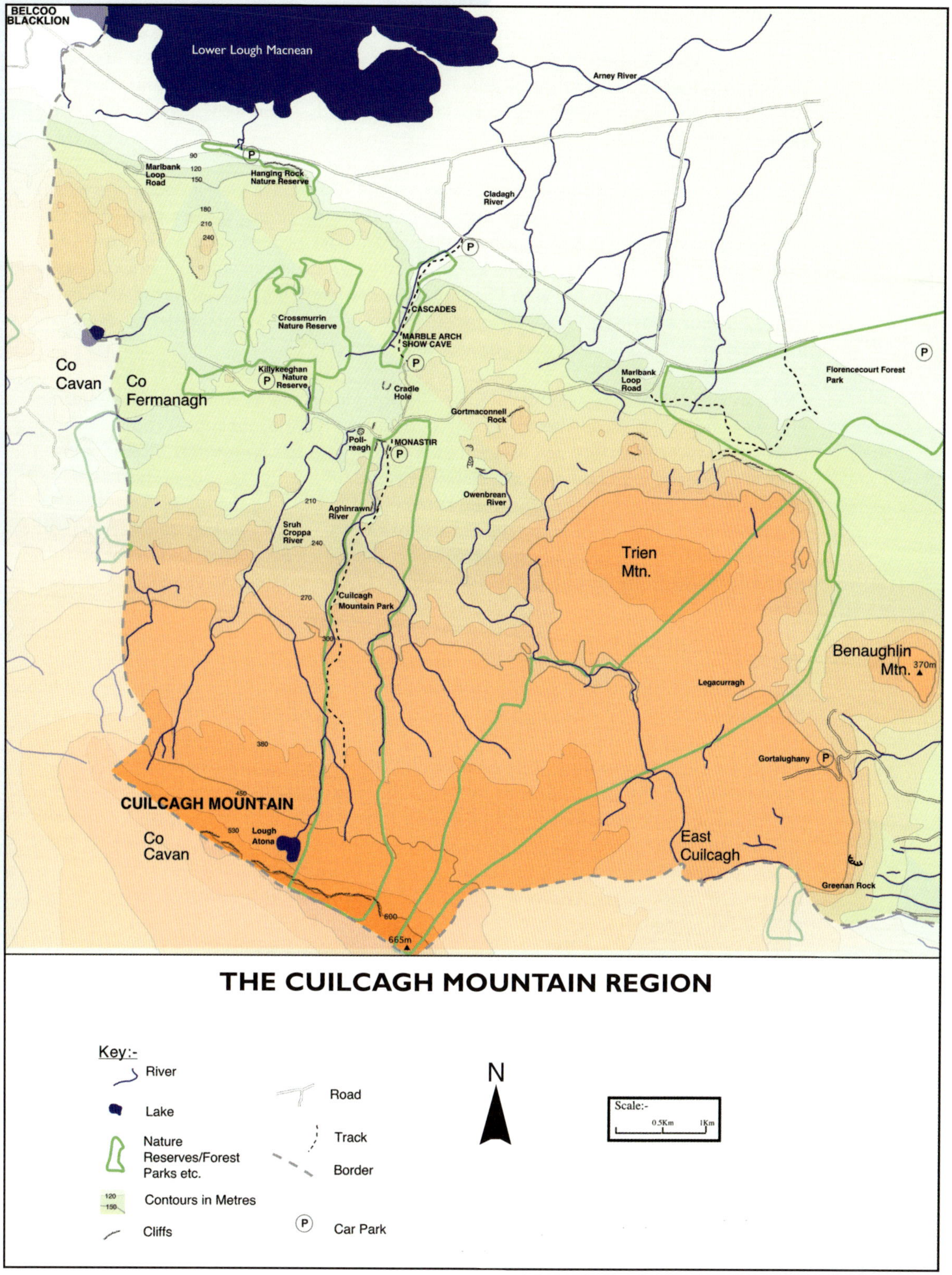

THE CUILCAGH MOUNTAIN REGION

The Caves and Limestone Scenery of the North of Ireland

Ancient earth movements shifted blocks of land one against another creating major breaks, most notably the Cuilcagh Fault which cuts east-west across the northern slopes of the mountain. Movements on a grander scale slightly tipped the once horizontal beds of rock so that, in general, they now dip slightly from the east down to the west and south. Later, volcanic action then pumped molten rock into the ready-made weakness of the Cuilcagh Fault, creating the Cuilcagh Dyke.

The thick beds of sandstone are seen clearly in the cliffs above Lough Atona and a little further to the west at the Cuilcagh Gap. Below the cliffs, slopes of dark grey shale screes are peppered with massive sandstone blocks which have broken away from above. For karst features, it is the wide terrace in the Dartry Limestone stretching from the east end of Cuilcagh, westward to the area known as the Marlbank, which is of interest. It is here that the major known caves are developed. East Cuilcagh is characterised by deep potholes, while the Marlbank can boast of fine river caves.

Our earlier textbook description of water flowing down the mountain and disappearing underground at the shale-limestone contact is, in reality, an over-simplification. In many places, the water does not sink as soon as it encounters the limestone. Instead, stream beds are covered with boulders and debris washed off the slopes above. This has the effect of guarding or protecting the limestone surface and preventing the water from entering the fissures immediately. In addition, the Cuilcagh Dyke running from east to west through the area acts in some places as a natural barrier to the water, influencing and diverting its flow over and through the limestone. It is just visible on the surface in a number of places (Fig. 50).

The faults and the dip of the beds have also influenced the flow of water so that the underground rivers do not necessarily follow the simple downhill routes you would expect by looking at the surface topography.

The major cave systems known at present all bring water northwards from Cuilcagh Mountain to the Macnean valley. However, water tracing experiments prove that the flow, unexpectedly, also tracks underground beneath the slopes of Cuilcagh from its east end to its west end, emerging at Shannon Pot in County Cavan. The deep, still waters of this spring are the source of the River Shannon which flows through Ireland to enter the sea at Limerick; it is not widely known that this, the longest river in Ireland, has its beginnings in a cave in Fermanagh.

Experiments on the chemistry of the water rising from Shannon Pot reveal that some of the water is ancient and has gathered and ponded underground for maybe thousands of years. It is possible that a very deep drowned cave system lies under the mountain.

Water tracing has also shown that the underground drainage in the Cuilcagh area is further complicated by the fact that some of the water from East Cuilcagh breaches the Cuilcagh Dyke to flow north and enter the Tullyhona and Cascades cave systems. It is possible to speculate that it was the glacial deepening of the Macnean valley which caused the change in patterns of flow (Figs.58 and 61).

The Caves and Limestone Scenery of the North of Ireland

Fig. 60 The karst and caves of the Cuilcagh Mountain region.

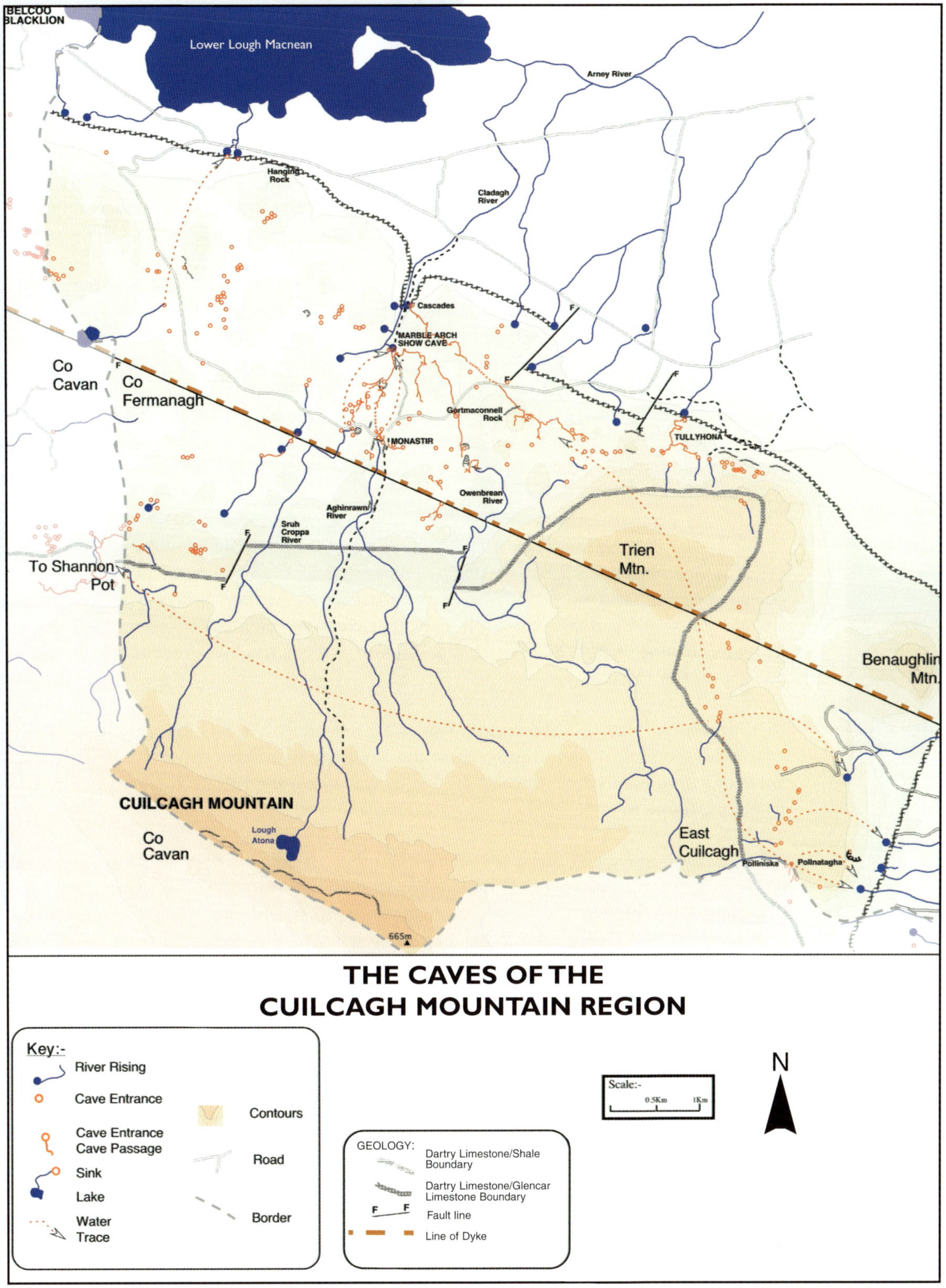

THE MARLBANK

Streams flow down the northern slopes of Cuilcagh and across its great expanse of bog. They flow onto the limestone and sink from view. This is where the Marlbank karst begins. From these vanishing points the cave-bearing Dartry limestone extends like a rocky shoulder across a three kilometre wide strip to the foot of the steep, wooded escarpment in the Macnean valley. Limestone pavement is the hall-mark of the area, sometimes exposed, sometimes barely concealed by a thin buttering of soil, and sometimes buried by glacial debris known as boulder clay. Knolls rise out of the green meadows, rushy hollows and hazel copses, their craggy white slopes reflecting brightly in sunlight. These hills, which started life as limey mud banks in an ancient tropical sea, now give the countryside its picturesque appearance (Figs. 43 and 44). Streams are absent from the surface landscape, the sound of flowing water is rarely heard here. Depressions, collapses and dry valleys hint that the underlying rock is riddled with subterranean waterways and cave passages. Sandstone erratics lie, dumped on the limestone, a long way from their parent rock on the upper slopes of Cuilcagh Mountain.

Fig. 61 Hanging Rock is a knoll which was sliced in half by moving ice. Now we can see what a knoll looks like on the inside. The steep white face has few bedding planes.

The surface karst scenery of the Marlbank can be appreciated simply by driving along the Scenic Loop. The road leads through the very heart of limestone country, winding beneath Gortmaconnell Rock, passing by Pollreagh doline, and, in some places, actually going over cave passages. At one point, the road crosses a small bridge over the Sruh Croppa River. Here, depending on the weather, the river bed may be completely dry, all its water flowing underground in cave passage or, if there has been recent rain, it may be flowing on the surface.

At Killykeeghan Nature Reserve, mid-way along the Loop, it is possible to take a short walk from the car park and see some of the karst features close up, particularly limestone pavement, its clints finely etched by rainwater and its grykes sheltering limestone flora.

As well as the pavement, there are numerous other glaciokarst sites in the area, all vividly illustrating the massive power of moving ice and its melt water. Above the major branches of the Marble Arch cave system, tucked into secluded corners in the landscape, are East and West Gorge converging into a dry valley, formed when torrents of icy water flowed over, instead of under, ground.

The spring line, the points at which underground water re-surfaces, can be seen along the Macnean valley, at the foot of the escarpment. The Florencecourt-Blacklion road runs along the base and crosses a number of these resurging rivers, the biggest being the Cladagh.

The Caves and Limestone Scenery of the North of Ireland

It is possible to speculate that the Fostra, a dry gorge next to Hanging Rock, was formed by glacial water torrenting from a melting ice edge. The Lough Macnean valley itself was modified by moving ice, its sides trimmed and its floor eroded. This gradual deepening of the valley would have resulted in a lowering of the water table and this in turn would have left areas in the underground drainage system high and dry, as Figure 11 illustrates.

The Marlbank's three major cave systems are Tullyhona, Cascades and Marble Arch. Some of their river sinks are impressive, with water rushing along sheer-sided gorges before disappearing at the base of 30 metre high cliffs. In other cases, the streams quietly vanish into their beds.

All these caves can be described as dendritic, which means that, like branches on a tree trunk, streams sink in different places on the hillside and come together in a main river underground. In most cases, it is not possible for cavers to make an expedition from sink to rising as sections of the passages are completely water-filled, lying drowned below local water tables, and dry routes are often blocked by areas of collapse or by complete sediment blockages.

The dimensions of the main passages vary from just about human-size to spacious tunnels up to ten metres wide by ten high. Most are active, echoing with the sound of water, as the rivers spill through the caverns. Sometimes, silence reigns when the flow is temporarily slowed in velvety black lakes. Sandstone boulders strewn along the passages gleam jet black with their coating of manganese oxide precipitated from the river water. Water pours and drips from the fissures in the roofs and walls leaving all manner of calcite decoration.

Fig. 62 The vertical walls of Monastir Gorge.

Fig. 63 A splendid array of curtains hang above the streamway in Tullyhona Cave.

Fig. 64 A caver swims through the inky blackness of Pollnagollum lake.

The Caves and Limestone Scenery of the North of Ireland

West of the Tullyhona and Cascades cave systems, the three rivers which feed the Marble Arch system drain an area of Cuilcagh mountain of approximately twenty seven square kilometres. They flow in separate cave passages and eventually join underground to form the Cladagh River. A section of this system has been developed as a show cave. It is possible to take a guided tour from an entrance close to the point where the Cladagh River exits the cave. A boat takes visitors upstream to the junction of the three rivers. The tour then continues on foot up one of the three active river passages.

The Cladagh surfaces just behind a natural limestone bridge known as the Marble Arch. In normal conditions, about one and a half cubic metres per second of water emerges, but this regularly rises to ten times the volume in flood. With one cubic metre of water weighing one tonne, the forces involved are terrific. This volume of water makes it the largest spring in Ireland. Confusingly, the Marble Arch is not marble but limestone. It was given that name in bygone days because of its smooth white marble-like appearance.

A perfect example of a collapse doline lies just behind the Marble Arch resurgence. It acts as a skylight into the underground river and can be viewed from the footpath which winds around its lip. In times of heavy rain, the sound of flowing water can be heard echoing out of it.

Further down the Cladagh Glen within the Marble Arch Nature Reserve, the textbook example of a resurgence is dashed by the sight of water rising from the Cascades system. Here, the water emerges from small, low bedding planes, having broken through the impervious layers in the Glencar Limestone to make cave passages in rock which, according to the book, should have none. For the caver, exploring this cave means moving through the Glencar Limestone in a maze of tiny, wet passages, until, in a flash of geological magic, a tight squeeze upwards leads into the spacious chambers within the Dartry Limestone.

The cascading waterfalls at the resurgence, which gave the system its name, are also a place to see tufa - a soft form of calcite encasing vegetation as it trickles out of the limestone. A more detailed description of how it forms is given in the next chapter (Fig. 94). Like all cave deposits, it is delicate, and should not be touched.

West of the Cladagh Glen, more streams sink into the limestone terrace above and emerge from the springs below but, as yet, no major cave systems have been found. Glacial debris chokes the dolines which are the potential entrances to cave passage. All underground exploration so far has met sumps: drowned water-filled passages. Hanging Rock Rising, concealed by dense ash woodland, is a major spring which, so far, has no explored open cave passage to its credit, (Fig. 22). Its sink and associated dolines, over a kilometre away and 100 metres above, close to the Marlbank Loop, can only be penetrated for a short distance.

Fig. 65 Close to the resurgence, on the far bank of the river a short distance downstream from the Marble Arch, the contact between the cave-bearing Dartry Limestone with the thinly bedded, water resistant Glencar Limestone beneath can clearly be seen. It vividly illustrates why the Cladagh River has been forced to come to the surface at this point.

The Caves and Limestone Scenery of the North of Ireland

Fig. 66 The Marble Arch Cave System.

The sediment floor which this flowstone originally formed across has been washed away leaving the canopy hanging in mid-air. Cobbles stuck in the underside of the calcite are all that remain of the floor.

The tips of these unusual stalactites formed in mud. Water has since removed the mud leaving these balls of calcite suspended above the cave floor.

Magnificent flowstone in Pollnagollum river passage.

The barely visible figure of the caver gives some idea of the grand scale of Giant's Hall.

The river flows underground at Pollasumera.

The cave plan shows the rivers flowing in separate underground passages before coming together to form the Cladagh River.

The Caves and Limestone Scenery of the North of Ireland

> ### ■ DATING CAVES
>
> It would be very satisfying if it were possible to date our cave systems accurately, and with advances in sediment dating, scientists are getting closer. Various dating methods are used including radiometric dating which determines the date of a deposit, based on the known decay rate of radioactive elements such as carbon-14, which occurs naturally in certain rocks. If, for example, we can find out the age of the calcite in a stalactite, we are then able to say that the cave in which it formed must be considerably older.
>
> Uranium series dating was carried out on speleothem samples gathered from sediment in Legnabrocky Way, in the Marble Arch System. The samples were fragments of flowstone broken by melt-water rivers during glacial times and preserved in the sediment as calcite pebbles. The pebbles were calculated to be at least 320,000 years old, which means that we are able to say that this particular cave was already well-developed approximately half a million years ago.

EAST CUILCAGH

East Cuilcagh, although part of the same Cuilcagh mountain mass, is different in character to the Marlbank for a number of reasons. A thick blanket of peat conceals most of the underlying limestone. Here the Dartry Limestone is not as pure, but is filled with black nodules of chert and instead of horizontal cave passages, it has deep potholes punched into its surface. Depressions in the bog, their presence indicated by stunted willow and rowan trees sheltering within them, and the distant sound of falling water, are often the only clues to hint at the presence of these geological traps! This area should be traversed with extreme care.

East Cuilcagh's wild setting can be appreciated from the Gortalughany car park which affords a panoramic view west, to the expanse of Cuilcagh, as well as eastwards to Upper Lough Erne. Rising out of the green and watery Erne lowlands, to the south-east, are the isolated, outlying limestone outcrops of Knockninny Hill and Molly Mountain. Although they share a geological make-up similar to Cuilcagh's, they cannot boast much in the way of karst landscape other than a few dolines, a couple of stream sinks, and small relict cave on the western slopes of Knockninny.

East Cuilcagh has twelve major deep potholes. The deepest is 70 metres from the entrance to the lowest point and none of them have much horizontal passage at the bottom. They are in a cherty zone, so different in character to the purer, thickly bedded, cave-friendly limestone further to the west. There are few stalactites to decorate their sombre muddy walls. Instead, debris from past floods festoons the walls with grass and twigs. The pot holes have all developed on major vertical fracture lines which criss-cross this limestone. The streams flowing off Cuilcagh have concentrated on widening these

Fig. 67 Benaughlin Mountain, a 370m high distinctive landmark at the edge of the East Cuilcagh area. The cherty Dartry Limestone forms steep cliffs on its northern and eastern sides. Perched precariously high in the cliffs are small, relict caves called the Brocky Caves.

weaknesses to create plunging shafts; however, the trickles we see today cannot possibly account for their size. Instead, the grand scale of the potholes can be attributed to the massive eroding power of abrasive, sediment-laden, glacial melt water. At the bottom of the potholes, the water cannot be followed much further by cavers. It disappears from human view into impenetrable bedding plane passages. Dye has been put into the water and traced to the risings, confirming the existance of a complex underground drainage pattern in which the water goes in three different directions - east, north and west. Exactly what happens inside the mountain may never be viewed by cave explorers and may still be debated long into the millenium!

Just as in the Marlbank, there are many glaciokarst features in this area. Figures 69 and 70 illustrate how some landforms along the limestone escarpment were created at the edge of the melting ice.

Fig. 68 Pollnatagha, on the Fermanagh/Cavan border is one of the deepest potholes in the area. Water cascades 50 metres to the floor of a chamber. From here, and unusually for East Cuilcagh, cave passages can be followed by pot-holers along tight vertical fissures to the bottom of yet another series of potholes which lead back up to the surface.

Fig.69 The grass-floored dry valley of Legacurragh comes to an abrupt stop at its western end. We can imagine how glacial melt water flowed out in torrents from the edge of the ice and carved the valley in the frozen ground.

Fig. 70 Tucked into the escarpment is Greenan Arch, another enigmatic feature which probably owes its origins to glacial melt water.

The Caves and Limestone Scenery of the North of Ireland

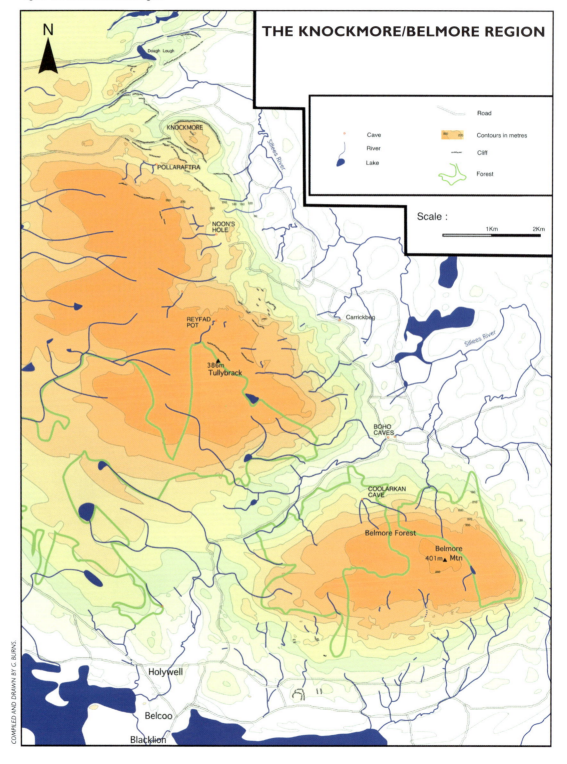

Fig. 71 The Knockmore/Belmore region.

■ FROM KNOCKMORE TO BELMORE MOUNTAIN

This area, lying north of Lough Macnean and west of Derrygonnelly, rivals Cuilcagh for surface and underground karst interest.

Even though its geological character is the same as Cuilcagh's, it also has some distinctive features of its own. It is slightly more difficult to describe as it is not one definite mountain mass like Cuilcagh. Instead, it is an extensive area of uplands, its highest points being Tullybrack Mountain, which just manages to rise above its surroundings to a height of 386 metres, and Belmore Mountain, 401 metres. Forestry plantations and expanses of bogland mask the contours of this undulating sandstone plateau which stretches from Big Dog in the north, through the Ballintempo Uplands, to Belmore Mountain in the south. Belmore is detached from the main block by the Boho Valley.

Just as in neighbouring Cuilcagh, sandstone cliffs can be seen as dark scars on the skyline. The thin beds of shale which underlie the sandstone step down to the limestone escarpment, with its terraces of Dartry Limestone exposed on the eastern and southern flanks.

North of Knockmore, the Doagh caves lie high and inaccessible in cliffs overlooking the lough. They are relict caves formed on the faulted contact between the limestone and sandstone. An ice edge and subsequent glacial erosion may have removed most of the original pre - or inter-glacial cave passage leaving only these short remnants.

Fig. 72 Steep cliffs line the edge of the limestone escarpment, the most impressive being Knockmore overlooking Derrygonnelly.

The Caves and Limestone Scenery of the North of Ireland

TULLYBRACK

The cliffs of Knockmore silhouette the skyline at the northern end of Tullybrack Mountain's limestone escarpment. The area boasts many of the features necessary to capture the imagination of a karst enthusiast; the underground sites in particular are of regional, national and international scientific importance.

Fig. 73 The limestone pavement in this area is some of the best developed in the north of Ireland.

The caves of this area are extensive and varied. They have all formed on the north-eastern slopes of Tullybrack Mountain. Their pattern of development has been influenced by a number of east-west faults which dissect the limestone. These faults are part of the Castle Archdale-Belhavel Fault Complex, which trends from Mayo to Scotland.

Of the many cave sites, the most notable are: Reyfad, one of the longest cave systems in Ireland; Noon's-Arch, which contains the deepest pothole in Ireland; and Pollaraftra, a splendid, fault-orientated river cave.

The Reyfad cave system lies under a peat covered terrace which gives little indication, other than the groups of dolines, of the spectacular cave system running below. Although nearly seven kilometres of known passage exist at present, it remains one of the unknowns in Fermanagh cave exploration as only a fraction of its potential has been realised. Great gaps exist on the survey between its explored cave passages and its resurgence at Carrickbeg near Boho Chapel.

The extensive underground network cannot be attributed solely to the action of one major river during one phase of development; in fact, nowadays, five small streams enter the cave and three of these lead via deep shafts to the huge tunnel-like passages 100 metres below the surface. The main cave has been geologically guided by two sets of fissures giving the cave its zigzag look. Strangely, as the survey illustrates (Fig. 75), a long straight passage heads south, taking the stream far under the sandstone cap of the Ballintempo area and in the opposite direction to its proven rising. This particular passage has a gloomy character as the sandstone cap above limits the seepage water needed to create calcite decoration.

The passages are packed with evidence of a long and complicated evolution. Sand and mud banks built by melting glacier water fill much of the cave. During the various phases of the Ice Age, the presence of an ice mass in the Sillees valley must have had a significant effect on the caves and underground drainage. Ice blocked the outlets, so raising underground water levels and possibly drowning parts of the caves. The subsequent melting of the ice plugs would have allowed water levels to drop again, so dynamically altering the build up or removal of sediment. These sediments contain valuable information about the changing environments of the past and represent some of the most extensive subterranean glacial deposits in the country. The ice deepening of the valley means that the resurgences, at the Dartry-Glencar contact, are perched high on the slopes above the valley floor.

The Caves and Limestone Scenery of the North of Ireland

Fig. 74 The karst and caves of the Knockmore/Belmore region.

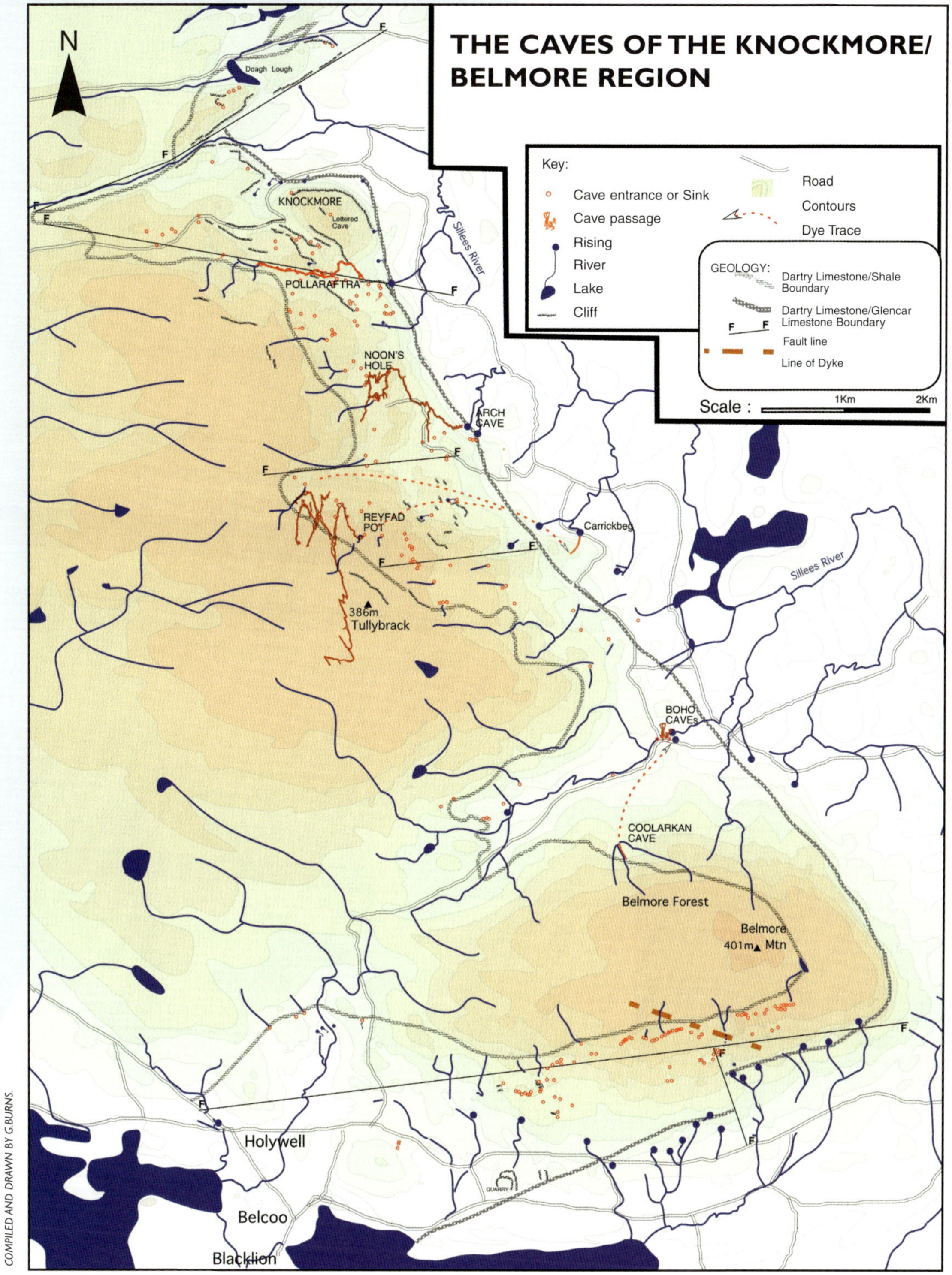

The Caves and Limestone Scenery of the North of Ireland

Fig. 75 Reyfad Cave System.

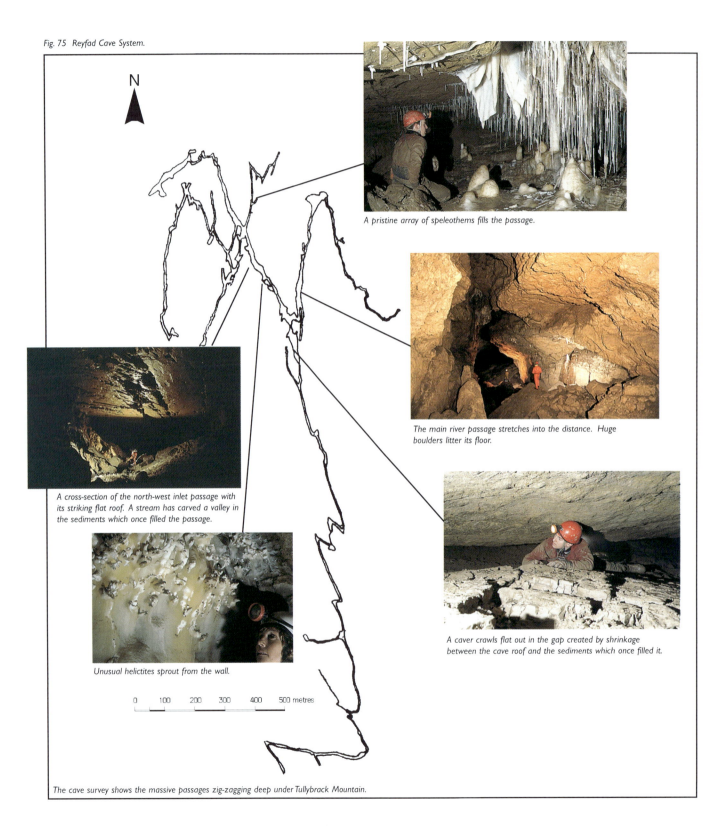

A pristine array of speleothems fills the passage.

The main river passage stretches into the distance. Huge boulders litter its floor.

A cross-section of the north-west inlet passage with its striking flat roof. A stream has carved a valley in the sediments which once filled the passage.

A caver crawls flat out in the gap created by shrinkage between the cave roof and the sediments which once filled it.

Unusual helictites sprout from the wall.

The cave survey shows the massive passages zig-zagging deep under Tullybrack Mountain.

The Caves and Limestone Scenery of the North of Ireland

North of Reyfad, a small clump of trees and a trickling stream is all that hints of the deepest single shaft in Ireland and possibly its finest: Noon's Hole. The Noon's-Arch system has a small water catchment of only three and a half square kilometres.

The water tumbles over the lip of the pothole, splashing off the walls and filling the shaft with spray. The walls are sculpted and fluted by the falling water. At the base of the pothole, 100 metres from the surface there is still the faintest glimmer of daylight. The water flows into pools and tight rifts, then through a water-filled passage and into a kilometre of magnificent open passage decorated with a great array of stalagmites. After long lakes and a further two sumps the river returns to daylight, emerging from a cave mouth appropriately named Arch because of it high vaulted roof. The river gushes out of darkness over a series of cascades and flows down to join the Sillees River in the valley below.

This cave river is one of the few in Fermanagh that can be followed by cave divers along its complete route, from daylight back to daylight, from sink to rising.

Fig. 76 Noon's Arch System.

Deep in the woodland the river emerges at Arch cave.

Noon's / Arch system. Daylight filters down the 100 metre entrance shaft, barely illuminating the abseiling caver as he nears the bottom of the pothole.

The river passage in Arch 2.

The cave explorer's journey from Noon's Hole to the Arch resurgence can be followed on this map of the underground passages.

The Caves and Limestone Scenery of the North of Ireland

Pollaraftra Cave is unique in Fermanagh in that it is guided for most of its two kilometre length by a major fault (A-B). As Figure 77 shows, this controlling fault is clearly visible, both on the surface and underground. The streams leave the sandstone and when they encounter the fault face they sink and follow it underground almost unswervingly to the rising. The inclined fault forms one wall of the cave giving it a triangular-shaped cross-section. Marks called slickensides have been gouged in the fault surface by the friction caused by ancient movement between the moving blocks. The walls of one chamber are covered with tiny beads of calcite precipitated onto the walls from the spray of falling water. This unusual speleothem is called botryoidal calcite or, more descriptively, cave popcorn.

It is almost possible for cavers to make a complete underground journey from sink to rising, but not quite. The river's final few underground metres go through passages too small for cavers to follow. Its rising is at the top of a series of picturesque waterfalls which descend through thickly wooded slopes.

Fig. 77

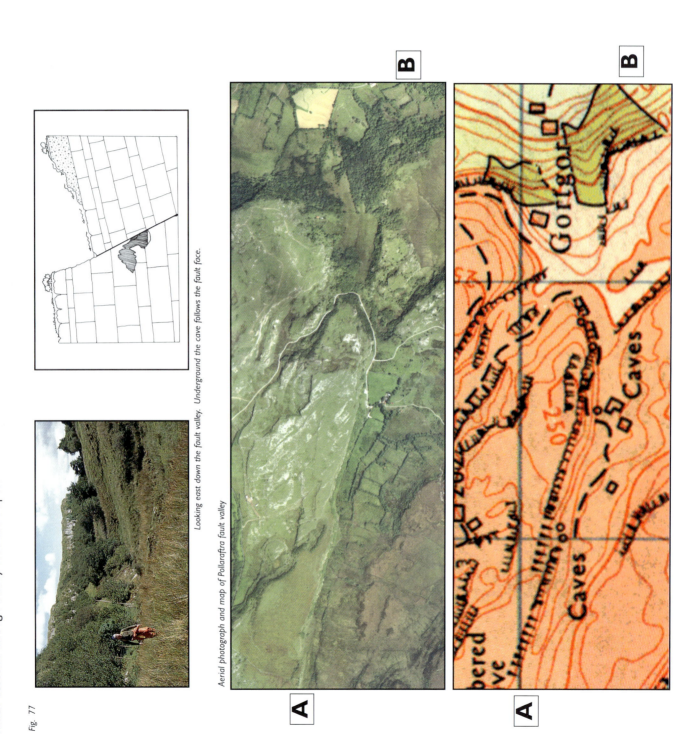

Looking east down the fault valley. Underground the cave follows the fault face.

Aerial photograph and map of Pollaraftra fault valley

The Caves and Limestone Scenery of the North of Ireland

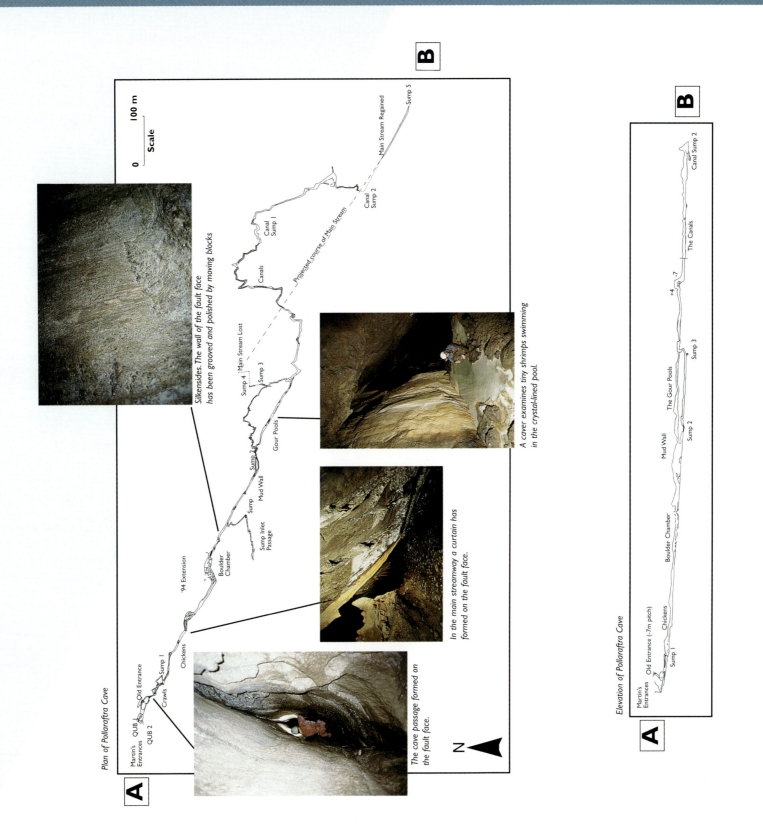

The Caves and Limestone Scenery of the North of Ireland

THE BOHO VALLEY

The scenic Boho valley divides Belmore Mountain from Ballintempo. It has two important cave sites, Coolarkan and Boho. Coolarkan is actually on the northern flanks of Belmore but drains into the Boho valley. It has a single, large, flat-roofed passage which heads dramatically into the hillside looking as if it will punch right through the mountain, only to end abruptly 100 metres beyond the entrance at an ancient boulder blockage. The small stream which today tumbles over the back wall of its entrance and flows through it, has been traced with dye to the complex underground drainage of the Boho caves.

Boho Cave, a classic maze cave system near the village after which it is named, has held local fascination for years. It is a natural labyrinth in a small block of heavily jointed chert-filled Dartry Limestone. The quarrying which modified it ceased in the 1950s. A glance at the survey shows the complicated lay-out of its two and a half kilometres of open cave. This amount of passage within such a small area of rock has come about because the volume of water flowing into it is too great for the size and number of outlets. A head of water builds up, it back-floods and water forces its way into every joint in an attempt to get through. A maze of passages is created.

An extensive water-filled phreatic zone is assumed to lie under the open cave passage. When it rains, the water bubbles up to flow in the higher open passages. At the uppermost passage levels, above the flood zone, there is a series of long abandoned, relict passages with fine speleothems.

The most impressive feature of this cave is the dynamic manner in which it floods. It has a large catchment of approximately 16 square kilometres which collects rain falling on the sandstone uplands and feeds it down through a number of lakes into the Boho valley. It takes time for a flood to build up and reach the cave; sometimes as much as half a day has passed, the sun is out and the rain forgotten. But in the cave, things are just starting to happen. The cave wheezes as air is forced out of cracks by the rush of on-coming water. The peaty brown torrent fills the normally dry passages to the roof and eventually erupts from the quarry entrance to surge along the dusty stream bed and down the normally dry waterfall.

Fig. 78 The entrance to Coolarkan cave is a collapse doline with a re-invading stream cascading down its back wall.

Fig. 79 Boho Cave.

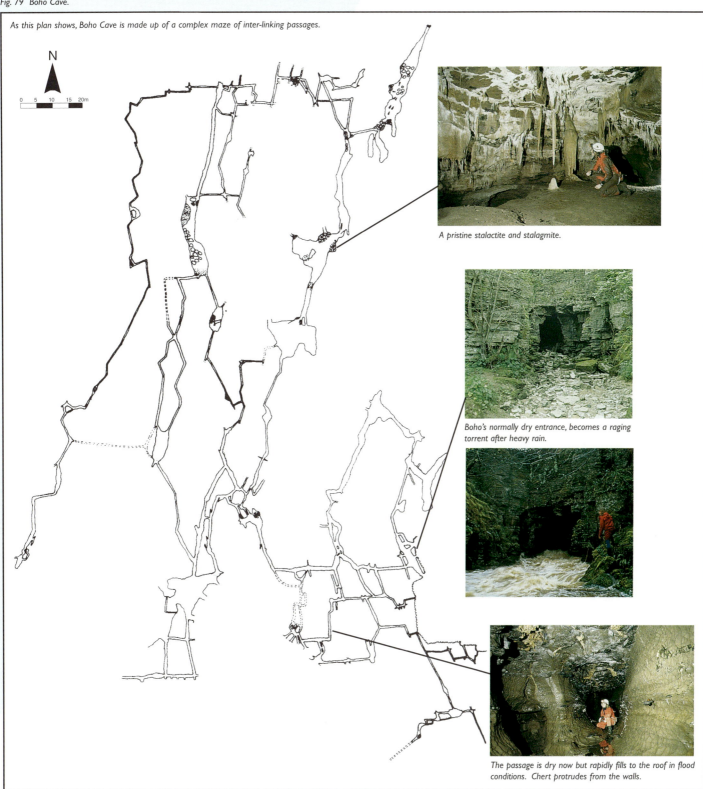

As this plan shows, Boho Cave is made up of a complex maze of inter-linking passages.

A pristine stalactite and stalagmite.

Boho's normally dry entrance, becomes a raging torrent after heavy rain.

The passage is dry now but rapidly fills to the roof in flood conditions. Chert protrudes from the walls.

BELMORE MOUNTAIN

Belmore Mountain is only 401 metres high, but it is prominent in the landscape as it is split from the main extent of the Ballintempo uplands to the north by the Boho valley, and descends to the shores of Lower Lough Macnean to the south. Its upper slopes are partly obscured with a dense covering of coniferous woodland.

Dartry limestone is exposed on its eastern and southern flanks beneath the cap of sandstone, but it contains no known major cave systems. This may be due to its complex geology. The limestone is dissected by the Belcoo Fault which runs east-west through the middle of it. A series of north-south faults have further split the rock and a volcanic dyke cuts through a corner.

The rock contains a significant proportion of the mineral dolomite which came about after the limestones were formed on the sea bed. Hot magnesium-rich water flowed up cracks and replaced the calcite with dolomite. This mineral gives the rock economic value as it is in demand for various agricultural and industrial purposes.

Though lacking known caves, Belmore has numerous surface karst features, such as dry valleys, which may be associated with the positions of past ice edges. There are also many springs, the most noteworthy being the Holywell rising which lies on the Belcoo fault (Fig. 106). The source of this water is not known although it has been suggested that it may track underground from Reyfad, eight kilometres to the north.

THE NATURAL BRIDGES

West of the Ballintempo Uplands, in the townland of Tullybelcoo, is a small outcrop of Dartry Limestone. The River Roogagh, which enters Lough Melvin at Garrison, has carved a short section of cave through bedding planes in the limestone where it makes contact with sandstone.

Nearby, in the low lying valley floor near Kilcoo, the County River has found a course through a small outcrop of limestone, forming a natural rock bridge with a span of eight metres, linking County Fermanagh with County Leitrim (Fig. 107). Folklore associated with the bridge is described in Chapter 7.

KILTIERNEY DEER PARK

An isolated and intriguing area of lowland limestone lies near Irvinestown on the northern side of Lower Lough Erne. A karst drainage system has developed within a small area of heavily faulted Early Carboniferous Ballyshannon Limestone, most of which lies within the walls of the farmland of Kiltierney Deer Park. A stream flows out of Parkhill Lough and sinks in a pool close by. In high water conditions, the stream flows down the normally dry river bed. A spring, known as the Holy Well, emerges on the north bank of the dry river bed and promptly re-sinks and joins the main underground flow. (Local folklore weaves tales of times when fountains of water actually spouted out of the ground!) The Parkhill Lough River reappears in a collapse doline and flows across the bottom of the depression, before disappearing again underground into Fiddler's Cave. The cave is short, low and muddy and ends at a sump after a short distance. The archaeology of this area is particularly special and is described in more detail in Chapter 7.

■ *FARDRUM AND ROOSKY TURLOUGHS*

Close to the southern shore of Lower Lough Erne, near Ely Lodge Forest, are two low lying karst lakes, known as turloughs. They are formed in depressions on the Ballyshannon Limestone and drain underground into the bedrock and refill as the groundwater level rises in periods of wet weather. During longer spells of dry weather the depressions may empty completely of water (Fig. 23).

Vanishing lakes such as these are a relatively common feature in the karst of County Clare but there are few clearly defined turloughs elsewhere in the karst of Britain or Ireland.

Fig. 80 Fardrum and Roosky Turloughs. These photographs were taken at different times. Roosky is empty. Its high water level can just be made out on the trees surrounding the 'lake'.

COUNTY ANTRIM

The white sea-cliffs, topped with a contrasting layer of black volcanic rock, give this stretch of the County Antrim coastline its distinctive character. Here, in these vertical faces, the Ulster White Limestone with its dark lumps of flint dotted through it, is very much in evidence. However, as the map shows, it does not appear extensively but instead snakes in narrow bands along the edge of the basalt-capped Antrim Plateau. Lying below the limestone in this area are Jurassic clays called the Lower Lias clays which are full of fossils.

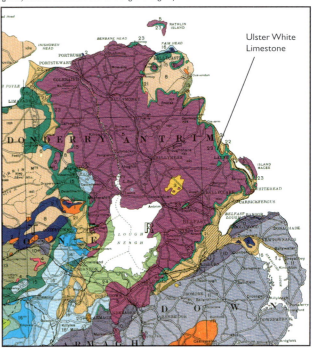

Fig. 81 The geology of County Antrim. The Ulster White Limestone (colour-coded green) snakes in narrow bands along the edge of the basalt.

■ THE LIMESTONE FOSSILS OF ANTRIM

Because the rocks in Antrim were formed during a different geological period, the fossils found in them are different from the Fermanagh fossils.

An electron microscope is needed to identify the Ulster White Limestone's micro-fossils of plankton. These are known as coccolithophores and are the main calcite source for this limestone. A photograph of them can be seen in Figure 47.

The common fossils of the Antrim limestone, which are visible with the naked eye, include belemnites, echinoids and sponges. Belemnites when seen in profile are bullet-shaped, and seen end-on, they appear circular. They are the internal skeleton of a creature similar to the present day cuttlefish or squid. Their name comes from the Greek for 'dart' and they were thought, in folk legend, to be 'thunder-bolts'.

Echinoids are fossil sea urchins and as their shape often resembles a heart, they have been called 'fairy hearts'. Sea urchin shells are commonly found on our beaches today. Fossil sponges found in the Ulster White Limestone are roughly cone-shaped, if viewed side-on but look circular, when found end-on.

The Lias Clay which lies under the Ulster White Limestone is fossil-rich. Fine examples of bivalves, crinoids and ammonites can be found.

The Caves and Limestone Scenery of the North of Ireland

In the geological sequence, there is a gap between the top of the limestone and the bottom of the basalt. This is called an uncomformity. The rocks which occupied this gap were removed by erosion before the lava spilled over to form the basalt cap. During this period of erosion, which took place approximately 60 million ago, karst features such as dolines and caves developed. They were filled with volcanic debris, then sealed under the basalt, and now can be seen clearly in the cliffs and valley walls (Fig. 24, and Fig. 82). These palaeokarst features are clear evidence that there was cave development in Antrim long before the start of volcanic action, and dispel the theory that the Ulster White Limestone only acquired karst properties after it was baked and hardened by outpourings of hot lava. The rock's hardness can instead be attributed to calcite recrystalization.

■ THE CAVES AND KARST SCENERY OF ANTRIM

Because the Ulster White Limestone outcrop is not extensive, its visible karst features are limited and often difficult to identify; however, sinks, gorges, dolines, caves and shafts can be found throughout the area. Some of these are described below.

Fig. 82 With just a little bit of imagination it is possible to make out the circular feature in the face of Magheramorne limestone quarry near Larne. It is possible that it was once a cave passage, later filled with sediment and finally covered by molten lava. Fossil karst features remind us that great changes have taken place in the landscape since cave passages were first carved.

A number of streams drain the east facing scarp of the Garron Plateau and sink close to the limestone-basalt contact. The joints and other fissures in the Ulster White Limestone act as the controls for the movement of water through it. The streams then resurge at the contact with the lower water resistant Lias Clay.

The most significant sinking river is the Black Burn which flows through an active cave system. It is special in that it is one of the most extensive, known karst systems developed in Cretaceous Limestones in the British Isles.

The Black Burn river cascades over the edge off the basalt plateau, then disappears underground at the base of the waterfall in the boulder-strewn floor of the gorge. The cracks and holes in the limestone which it sinks into, are crammed with tree trunks and other flood debris swept down from above. The cave has developed as a series of shafts which lead to horizontal tube-shaped passages. Its scallop-marked walls are a dazzling pure white, flecked with protruding bands of insoluble glossy dark flints.

Fig. 83 A waterfall pours off the basalt and sinks into the white limestone below.

The passages have been explored for 400 metres, by only one person, a cave diver, who has reached this point by swimming through the water-filled sumps using underwater breathing equipment. Only 140 metres have been comprehensively mapped and investigated because the passages are, for most of the year, partly or completely full of water. The water going into Black Burn Cave has been dye traced and proven to resurge at the limestone-clay contact, three and a half kilometres to the north-east at Garron Point. The resurging river must be one of Ireland's shortest surface rivers as it measures just 70 metres from spring to sea. During periods of wet weather the cave system overflows and the Black Burn reverts to being a normal river and runs on the surface down to the sea near Carnlough.

The Caves and Limestone Scenery of the North of Ireland

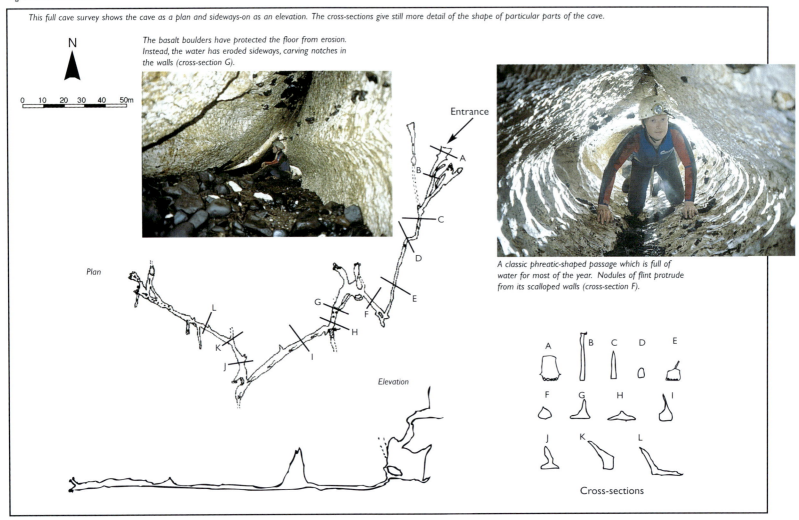

Fig. 84 Black Burn Cave.

This full cave survey shows the cave as a plan and sideways-on as an elevation. The cross-sections give still more detail of the shape of particular parts of the cave.

The basalt boulders have protected the floor from erosion. Instead, the water has eroded sideways, carving notches in the walls (cross-section G).

A classic phreatic-shaped passage which is full of water for most of the year. Nodules of flint protrude from its scalloped walls (cross-section F).

Fig. 85 Loughaveema, also known as the Vanishing Lake, on the road from Cushendun to Ballycastle. It cannot claim to be a true turlough.

Just behind the Garron Point rising is Garron Lough, which, although lying on limestone, is not a true turlough like those near Ely in Fermanagh, as some movement of water takes place via surface streams.

Near Tievebulliagh, west of Cushendall, in an area close to the escarpment line at the basalt-limestone contact is a series of solution dolines and sinks.

Within a small area of limestone in an area called Linford, west of Carncastle, some karst features can just about be recognised, including a group of well developed dolines and also active stream sinks and a dry valley. The dolines are developed along definite lines presumably associated with underground flow. The archaeology of this area is described in Chapter 7.

The Caves and Limestone Scenery of the North of Ireland

Along the Antrim coast there are a number of superb sea caves. Rathlin Island off the north Antrim coast can also boast of caves in the limestone along its southern shore. It is sometimes difficult to tell if these caves have been carved entirely by the dynamic power of the sea, or if they also owe their development in part to the dissolving of limestone by fresh water.

LARRYBANE CAVES

Not far from the Giant's Causeway, punched in the base of 30 metre high cliffs of Ulster White Limestone, are the caves of Larrybane Bay. There is no evidence to suggest that they were dissolved by water like 'true' limestone caves. Instead it is probable that they were eroded by the sea, even though they now lie some seven metres above the present sea level.

Fig. 86 Carrivemurphy Pot. In the sides of some of the valleys are shafts, some as deep as 20 metres and some with calcite deposits. These appear to be relict caves; the last remnants of cave systems left behind after ice scoured and deepened the valleys.

At the end of the 19th century, Martel, the eminent French cave scientist, visited and described them. He was particularly impressed by the striking development of tufa, which must have started to develop after the sea level dropped and the caves were left high and dry. Today, calcium carbonate-rich water still seeps from the cracks and encrusts the lime-loving mosses and liverworts growing at the entrance. Martel named the cave with the most extensive development of tufa, Larrybane Stalactite Cave, and, as the photograph shows, great pillars of tufa almost block its entrance. Inside, more tufa stalactites hang from the roof and drip and pour onto the floor creating stalagmites and gour pools. In the more dimly lit recesses some of the stalactites have a strange but definite slant towards the entrance; a probable explanation for this is that the vegetation prefers to grow where there is most daylight and so more tufa builds up on the brighter side.

Tufa is not just confined to the caves. High on the cliffs, at various points along the bay, cascades and stalactites of tufa 'pour' like frozen waterfalls from seepage points to the beach below.

Fig. 87 Larrybane Cave with massive tufa covered stalactites and stalagmites almost blocking its entrance.

Sharp eyes may also spot a miniature feature of the Larrybane Caves. Inside, on the upper parts of the shadowy cave walls, and again angled towards the light, are small patches of tube-like holes seemingly drilled into the rock surface, measuring no more than a centimetre in diameter and several centimetres in depth. Once again, plant growth is the explanation. Organic acids in algae growing in the dim light on the walls dissolve the rock to form this unusual **photokarren** feature, also known as biokarst or **phytokarst**. Just why the algae are able to excavate such deep tubes may be explained by the fact that the Ulster White Limestone is, as its name suggests, white and therefore highly reflective. Because of this, more life-giving light shines into the cavities and onto the algae, so allowing them to continue to grow and penetrate deeper.

The Caves and Limestone Scenery of the North of Ireland

COUNTY ARMAGH

Near Armagh City, tucked away in fields in the townland of Drumarg, is a small isolated outcrop of Carboniferous Limestone, locally known as the Sheep Walk. Within it, is a small cave system consisting of a network of short interlinked passages. Hidden by bushes, the three entrances to the system lie along the face of the limestone bluff. The caves penetrate underground for a distance of approximately 10 metres. They are completely dry, relict passages. The water that carved their domed roofs has long since disappeared. Webs of cave spiders festoon the walls, and foxes or badgers make use of the convenient dry shelter. In the field above, it is just possible to make out a doline which hints at the presence of passages running beneath.

COUNTY TYRONE

■ BIN MOUNTAIN

Near Drumquin, Lough Lee lies in a hollow on the top of a hill, south-east of Bin Mountain in a small area of Early Carboniferous Ballyshannon Limestone. It has no surface stream draining out of it. The water originally escaped down a sinkhole beside the lake and rose in the valley below at Rapid Spring. The sink and spring have since been extensively modified to supply the local water mains.

Close by Lough Lee, in a townland called Binnawooda, an unusual form of karst has developed in a type of sandstone which has a high calcium carbonate content. Here, a hillside terrace is puckered with dolines. A number of small streams sink into them. About 20 metres down the slope, is a spring called the Fairy Water. This site is of particular interest as it is the only known example of karstified sandstone within the north of Ireland.

■ SLIEVE BEAGH

This forested area straddles the Fermanagh-Tyrone-Monaghan border. Large areas of usually cave bearing Dartry Limestone forms outcrops on the western flanks of the sandstone and shale-capped mountain. The geology would suggest that there is good potential for karst and cave development; however, this is not the case. This may be due to the number of dykes dissecting the limestone which have inhibited the development of karst features. The only karst site of any note is a closed depression with a number of streams sinking into its floor.

Fig. 88 One of the relict cave passages at Drumarg County Armagh.

The Caves and Limestone Scenery of the North of Ireland

Beneath our Feet

Life above ground and life in darkness

ABOVE GROUND

There is no simple answer to the question, "what lives on the limestone?" The textbook answer would be "meadows rich in lime-loving herbs". Nature, however, doesn't read textbooks and things are rarely as straightforward. Instead, the answer has to be "it depends". It depends on the type and depth of soil on top of the limestone, it depends on how well it is drained, how steep the ground is and it depends on how much humans have interfered.

The best way to consider all these variations is to divide our surface karst scenery into a few distinct habitats: limestone smothered with bog, limestone grassland, limestone smeared with boulder clay, limestone cliffs and finally, woodlands. All of them have been shaped, to a greater or lesser extent, by the hand of man.

■ BOGLAND

Extensive areas of our limestone are concealed by bog. Here is perhaps the most striking example of human influence on the types of plant and animal life found on limestone. To understand why this is, we have to go right back in time, virtually to the end of the last Ice Age. When the ice finally released its grip of this part of Ireland, around 12,000 years ago, mosses and grasses were quick to recolonise, cushioning the rocks and softening the stark landscape. Giant Irish Deer, with antlers spanning a magnificent three metres, grazed a landscape of lush grasses and small willow and juniper trees. A sudden return to frozen conditions drove the deer to extinction but gradually temperatures increased and trees became established, until the whole landscape was thickly wooded, from mountain top to lake shore. With the dense forest cover came bears, boar, foxes and hare and with this plentiful food, shelter and fresh water, some 9,000 years ago came our Stone Age ancestors. At first, these people roamed the forest making virtually no impact on the landscape. Over the thousand or so years that followed, there was a gradual trend to a wetter and colder climate. Then the first farmers arrived. They cleared the forest for cultivation and livestock. The soils were laid bare, open to the leaching effect of the rain. Crops took away key minerals. Soils became acidic and waterlogged; conditions were ideal for the development of bog.

Fig. 89 The wild expanse of bog smothers the limestone on East Cuilcagh.

We must now ask the obvious question. If bogs rely on waterlogging and acidity to accumulate, how then can they possibly thrive on limestone? Limestone is, after all, quite the opposite; alkaline and criss-crossed with natural drains. The answer is that certain mosses can grow directly on the bare limestone, building up a bed of humus. Acid loving plants, like heather, grow on the accumulating cushion. The roots of these colonising plants make sure that they avoid direct contact with the limestone; instead, they draw sustenance from the humus. And so the bog growth is instigated.

Limestone bogs today harbour a wealth of acid-loving plants, like bog cotton, heather and the bright yellow bog asphodel, once used in place of saffron for medicine and dyeing. Insectivorous butterwort and sundew cleverly make up for the lack of nutrients in the peat

The Caves and Limestone Scenery of the North of Ireland

by trapping and digesting unlucky, visiting bugs. These bogs are an important habitat for birdlife too: snipe, golden plover and curlew, with their mottled brown plumage, are well camouflaged amongst the woody stems of the heather. For hundreds of years, peat has been dug for fuel, painstaking, back-breaking work. This human activity left little impact on the habitat other than the creation of turf banks; spade marks in the faces of the cut banks testify to the hard labour. Nowadays, the turf is cut by machines which churn across the bog, slicing into the peat and extruding it as long sausages. They can extract in days what took months to cut by hand. The bog can be radically altered overnight, including the drainage from it into the caves beneath. This important habitat, often viewed as Europe's 'rainforest,' once damaged often never recovers.

■ THE LIMESTONE GRASSLANDS

The terraces, rocky knolls and bare pavements of limestone which typify karst scenery support a special range of flora. Within this area are the herb-rich grasslands. Dry soils support a wealth of lime-loving plants which grow in profusion where the meadows have been traditionally farmed. Confusingly, the limestone soils are actually acidic. Heavy rainfall easily washes the lime away through the cracks in the rock. However, the lime-seeking roots have no trouble in penetrating the thin soils to reach the limestone beneath.

Fig.90 Thyme festoons the limestone, filling the air with scent on a warm day.

Today, it is careful management of the grasslands which ensures that the herb flora is maintained. Grazing by sheep or cattle is essential. However, the level of grazing has to be right; too little allows tall grasses to shade out the small herbs while heavy grazing prevents flowering and seeding and so the flora die out. Bare ground allows less desirable woody species to invade and establish themselves, so further reducing the plant diversity. The use of modern intensive farming methods on limestone also reduces diversity. The application of slurry and fertilisers forces grasses to grow rapidly, choking the natural herbs and preventing them from flowering.

But if the balance of management is just right, the reward can be stunning, even though the limestone hills of Fermanagh are 200 metres above sea level and growing conditions are harsher than, for example, limestone grassland at sea level. Pink thyme carpets the rock and bird's foot trefoil splashes yellow across the ground. Lady's bedstraw grows beside delicate harebell and mountain everlasting; names as colourful as their appearance. In the past, many of these plants such as eyebright, kidney vertch and milkwort were valued for their medicinal properties, their names still describe their herbal uses over the years.

Blue moor grass thrives on the grasslands of Fermanagh and nowhere else in the north of Ireland. Some species of orchids are common too on the limestone. The rare, dense flowered orchid has been found growing on Belmore Mountain.

The Caves and Limestone Scenery of the North of Ireland

Limestone pavement is a particularly special habitat. Within the clefts and deep humid fissures of these natural rock gardens there is a unique micro-climate which supports a distinctive range of plant species. Ferns such as hart's tongue and elegant wall rue find shelter in the crevices.

The herb-rich grassland provides the food and breeding sites for many insects. Butterflies, such as the common blue and peacock, lay their eggs on the bird's foot trefoil.

The Fermanagh limestone is home to the only known breeding colony in the north of Ireland of the small blue butterfly. The caterpillars feed on the yellow kidney vetch.

Growing in secluded dolines, around pot holes and on steeper hillsides, are clumps of woody hazel scrub. They only survive where they are safe from the hungry mouths of grazing animals. The hazel nuts are an important food for field mice and squirrels. Foxes and badgers often live on the edges of these hazel thickets, hunting for food in the fields around, while birds such as the wheatear are a common sight.

■ LIMESTONE UNDER GLACIAL DEPOSITS

In places where deposits of boulder clay were dumped by ice on top of the limestone, drainage is poor. The cracks in the rock are sealed by the sticky deposits. Limestone can no longer act like limestone and a different flora occurs. Swards of coarse, acid grassland dominate, with damp loving rush and vibrant yellow iris.

■ GORGES AND CLIFFS

On the cliffs and sides of gorges, vegetation manages to cling to the steep faces, searching out foot-holes in the cracks and clefts. Often, yew and juniper embrace the vertical rock, spreading their branches like arms across the walls, making a stark contrast against the white walls. Ivy snakes up the faces, creating shelter for nesting birds.

These inaccessible cliffs are often used as nesting sites by peregrine falcons; the safe, rocky remoteness is ideal for rearing young.

The mobile scree slopes beneath these cliffs also form distinct habitats. Amazingly, the jumbles of bare, sharp rocks provide just the right conditions for plants like mountain avens. Here, too, the bright yellow Welsh poppy thrives.

Fig. 91 Hart's tongue fern thrives in the damp warmth of a gryke.

Fig. 92 The small blue butterfly.

Fig. 93 Hazel copse, a haven for wildlife.

Fig. 94 Tufa on moss at a spring rising in the Cladagh Glen.

■ TUFA – PETRIFYING SPRINGS

Tufa appears where spring water emerges from underground. It is a crumbly, white, calcite deposit which precipitates from the carbonate-laden water. The exact nature of this precipitation is the subject of debate. It is thought to be due to a combination of the removal of carbon dioxide by living plants and the loss of carbon dioxide to the air as the calcium carbonate-rich water flows out over the rough surface of the vegetation.
A particular range of plants thrive at these so called petrifying springs, including some less common mosses and liverworts.
Deposits of tufa are infrequent and easily damaged.

■ WOODLANDS

Patches of remnant native limestone woodland survive, generally on steeper slopes where difficult access has, over hundreds of years, restricted clearance and halted agricultural improvement. Soils are thin and unstable too, where slopes are steep. Ash trees dominate on limestone as they favour its alkaline conditions.

Ash is late to come into leaf and its leaves are small. As a result, more life-giving sunlight is able to penetrate through the branches. In spring, the ground comes alight with woodland flora: bluebell, wood anemone, lesser celandine, wood sorrel and of course, primrose.

Hanging Rock Nature Reserve in Fermanagh is one of the finest native ash woodlands in the north of Ireland, (Fig. 61). It is believed that the great variety of lichens growing within it indicate that there has been woodland cover here since ancient times. In spite of its age there are few very old trees growing here as the steep mobile slopes beneath the cliffs can only support young trees. As soon as the ash become too big and heavy, they topple, leaving space and daylight for young ash to fill.

The rarest of all Irish mammals, the pine-marten, with its rich chocolate brown fur, creamy white chest and bushy tail, is a secretive creature, but it has been seen foraging in the limestone woodlands of Fermanagh. They are often known as marten cats and, quite possibly, are the cats referred to in Irish place names. So the cave on the Marlbank, known as Cat's Hole may owe its name to a 'marten cat' which once made its home in the shady cave entrance.

> ### ■ THE SPECIAL VEGETATION OF TURLOUGHS
>
> These disappearing lakes, set in the lowland limestone, have concentric rings of vegetation which indicate the different water levels. Hazel and blackthorn grows around the edge. They can only put up with a certain amount of water logging. Below this, is a ring of dense mixed herbs which then overlap with another of dense mosses. Finally, where there is a residue of water in the middle, there is a patch of fen-type vegetation.
>
> Some violets are particularly at home in turloughs; the Ely turloughs shelter the rare Fen Violet, with its pale blue rounded petals and slim pointed leaves.

LIFE IN DARKNESS

Caves are dark places. Deep underground, beyond the reach of natural light, the blackness is complete. It is difficult to imagine that any living creature can possibly exist in such a hostile environment. But exist they do and the study of them is known as **bio-speleology**.

Our caves are still recovering from the devastating effects of glaciation. Cave communities are small. Their survival is a delicate balance, easily upset if their environment is disturbed. In hotter climates, caves positively bristle with all manner of life. Some of these weird and wonderful cave-dwellers have adapted in bizarre ways over the hundreds of thousands of years of uninterrupted cave habitation.

What brings life to these crevices, cracks and caves? To answer this, we must first consider what any living thing requires to exist. The essential requirement is a source of energy for growth. Light is the energy giver for plants. Animals consume plant material or other animals, while the lowest forms of life, like bacteria and fungi, require organic material or minerals. Of course, none of these forms of life exist in isolation; from bacteria to mammals, from lower to higher plants, the plant and animal communities are interdependent. One creature's waste or body will become another's banquet.

Other factors, like temperature and protection from predators, will also influence the niches occupied by various forms of life. This chapter will delve into the depths, and peer into the nooks and crannies to investigate some of our cave-dwelling species and the lifestyles they adopt to exist.

In spite of the perception that they are hostile environments, parts of caves can actually offer shelter and a stable climate. The average temperature in the caves of Ireland is a constant nine degrees centigrade, regardless of whether there is a scorching sun shining or a winter blizzard raging. The temperature is naturally more variable near the entrance, becoming increasingly stable further underground. Flowing water influences the cave climate, and major draughts blowing through cave passages will also act like the rivers, pulling the surface heat or cold along with them. In any cave, therefore, there are a number of distinct micro-climates, particular conditions of temperature, humidity and air movement within a small area. The most stable micro-climates are in the low energy areas of the cave, places away from the draughts and flowing water.

An important point which we should bear in mind, is that we are big and bulky; gigantic by cave bug standards. The definition of a cave is a space big enough to be entered by humans; we can only visit a tiny percentage of the total underworld habitable by other life. Think of all the bedding planes, fissures and cracks, the **mesocaverns** and **microcaverns**, we can't get into. We are excluded from the very places most likely to shelter the specialised creatures. Cave biologists can only look at the creatures they find in cave passages and make intelligent guesses about what may lurk, tantalisingly, beyond their reach.

It is interesting to speculate on the amount of living space available for smaller creatures. An estimation of the amount of passages explored by cavers in the north of Ireland stands at about 35 kilometres. It would not be unreasonable to guess that there must be thousands and thousands more, penetrable by the likes of an insect. Just two orders of magnitude would put this at a length of 3,500 kilometres of living space. Imagine what this means, not only in terms of potential numbers of creatures but also in terms of brand-new species of cave creatures. There they may be, blissfully living their lives away in some desirable location, a bedding plane perhaps, with just the right micro-climate and food supply.

The box below lists the different ways of describing cave-dwelling species, depending on how long they stay in caves and whether they have actually changed appearance because they are permanent cave-dwellers. However, we will simplify matters and look at:

- species which live in cave entrances. This area is sometimes known as 'the threshold', or more tantalisingly, 'the twilight zone'

- cave visitors which use caves as temporary homes

- cave-dwellers which are able to live in total darkness for the whole of their lives.

The Caves and Limestone Scenery of the North of Ireland

■ CAVE DWELLING CREATURES CLASSIFIED

TROGLOXENE:
a species that occurs in caves but does not complete its life cycle there.

CAVERNICOLE:
a species that lives in a cave habitat and can complete its life cycle there. It may have changed its appearance i.e. it may lose its pigment, but if it returns to the surface it regains its former appearance.

TROGLOBITE:
a cavernicole that undergoes major bodily changes such as loss of eyes which suggest that it has undergone a long history of cave habitation.

TROGLOPHILE:
a cavernicole that completes its life cycle in non-cave habitats as well as cave habitats.

TROGLODYTE:
a human cave dweller. There is no evidence to suggest that there were troglodytes in any of the caves in the north of Ireland. Pot holers are the closest we get to troglodytes today!

■ FIRST THE ENTRANCE, THEN THE TWILIGHT ZONE

If we venture underground into the world of eternal darkness, we first must pass through the twilight zone of the cave. Here, light can penetrate a certain distance, depending on the size and aspect of the entrance. This is an important habitat for both plants and animals. It is sheltered from the wind and rain of the surface and although there is no direct sunlight, there is still enough for certain green plants to photosynthesise, in other words, to process light into life-giving energy.

Shade loving plants, like ivy and herb robert, grow at the entrance. The further we go in, the darker it gets and so the vegetation changes to plants which thrive in a damp habitat and are increasingly tolerant of shade.

Then, deeper in, ferns are replaced by the less demanding mosses and liverworts. Further still and lichens colonise the rock surfaces on the edge of perpetual night. Finally, the glimmer of light is so faint, only green algae and blue-green algae can exist. These single celled plants are able to live at low light intensities as they make the best possible use of the light they absorb. Unlike the higher, more complex plants, they do not have to use energy to make stems and leaves.

Fig. 95 Just a hint of daylight shining onto the wall of cave allows this green algae to survive.

The Caves and Limestone Scenery of the North of Ireland

Many animals make use of the entrance of a cave as a perfect shelter and safe home, sometimes temporary, sometimes permanent. There is evidence that, as recently as 7,000 years ago, brown bears (*Ursus arctos*) roamed this land and made use of caves. Bear skulls of a number of adults and cubs were found by cavers in the late 1990s, in a cave in the neighbouring county of Leitrim.

In the past, rock doves commonly nested in cave entrances. This is confirmed by the fact that many Irish caves are called Pollnagollum from the Irish, *poll nag Colm*, cave of the doves.

Nowadays, badgers are quick to recognise the potential of a dry cave passage as a ready-made sett. The cave is warm and its earth floor is rich in earthworms. A trail of grass dragged into the darkness is enough to indicate that they have their home within; the grass serves as bedding. Foxes too, may happily take up residence in dry caves. Again, cave names suggest their presence, Pollabrock (*poll an bhroic*), Badger Cave and Pollnamadrarua (*poll na madra ruaidh*), Hole of the Red Dog (fox).

Contrary to popular horror fiction, caves are not teeming with brown rats. The only time rats are found near them is when they have been attracted there by the promise of food. The irresponsible dumping of rubbish and organic waste is the problem. The humans are the culprits, not the rats.

In the twilight zone, there are the common woodland floor inhabitants like woodlice and millipedes, which are here because the cave floor is rich in organic material. However, there are some insects which seem to favour cave thresholds in particular, like species of caddis fly, mosquito (a non-biting one) and the herald moth and tissue moth. It seems that they can find their ideal micro-climate; a certain crack in the rock a certain distance from the entrance may have exactly the right temperature for them, either to hibernate or live actively. They often occupy exactly the same area of cave each year for the same length of time before leaving to resume life outside. In the process of moving in and out of the cave, they may find themselves prey to cave spiders.

The Caves and Limestone Scenery of the North of Ireland

■ THE CREATURES WE ARE MOST LIKELY TO SEE UNDERGROUND

COMMON CAVE SPIDERS

The larger and the lesser cave spiders (*Meta menardi* & *Meta merianae*) spin sticky orb-webs. In the dark, the spiders locate their prey by vibrations of the web. As well as trapping the flying insects, the webs also snare crawling bugs like beetles and millipedes. The spiders store their eggs in white silk, balloon-shaped egg sacs (the large cave spider's egg sac measures 25 mm long by 20 mm wide) which they suspend from the cave roof. When hatched, the tiny babies scuttle into minuscule cracks for safety.

The lesser cave spider favours existence within the reach of daylight while the large prefers to reside just beyond the limit of light penetration.

THE HERALD MOTH

Scoliopteryx libatrix seems to find caves the perfect place to hibernate. Compared to shelter above ground, a cave offers a constant temperature with no fluctuations in moisture. The caterpillars of the herald moth feed outside on willow and poplar but soon after emerging from the chrysalis, they head for the twilight zone of caves (or its man-made equivalent). The herald moth settles on the cave walls or roof, folding its deep orange wings, with their white and crimson markings, over its back. It has been suggested that, while underground, the females undergo a period of suspended development necessary if they are to produce eggs.

■ TEMPORARY DWELLERS OF THE DARK

Some cave visitors penetrate deeper into the caves. They are happy in total darkness but still leave the cave to forage for food. Perhaps the most well-known of these are bats.

Bats are quite choosy about where they live. They don't stay in the same place all the year round. What may seem haphazard patterns of roosting are, in fact, the result of careful selection according to their particular requirements at the time of the year. So a cave may have a bat colony or a solitary bat living in it one month and none the following.

During the summer months, some caves serve as nursery roosts for colonies of females and their young. Other caves are favoured as hibernation sites; the crevices in the rocks and cracks between boulders provide shelter and a stable temperature, essential if the bat is to survive the winter.

Bats use echo-location to navigate. They emit high-pitched noises which bounce off the surroundings. With this technique, they can make their way out of the cave and catch insects. They may 'turn off' their sonar equipment if they are flying in familiar territory. This explains why they may bump into an obstacle if it is not usually there. They quickly signal to following companions, so that they too don't collide.

Bats eat 50% of their body weight in insects each night. The guano beneath a bat roost is the fast food restaurant of the cave world. It is the basis for a food chain and crawls with feeding diners.

Fig. 96 A large cave spider (*Meta menardi*) waits for its meal to come calling. The egg cases of the cave spider may contain up to 500 bright yellow eggs.

Fig. 97 During hibernation, herald moths become very torpid; beads of moisture glisten on their wings. In this state they are vulnerable to predators.

Fig. 98 Daubenton's or water bats huddle together clinging to the roof of a river cave.

■ OUR CAVE-DWELLING BATS

Because bats favour deep cracks and crevices, it is difficult to say for sure whether a certain bat does or does not roost in caves. Often the only way to identify a bat or guess that it may be around is by the shape of its droppings.

Of the eight species of bat found in Ireland, six species (Pipistrelle, Natterer's, Daubenton's, Whiskered, Lesser Horseshoe and Brown Long Eared) have been found in caves.

Daubentons choose river caves. They swoop low over the water to catch the flies just above the surface. They particularly favour roosts in crevices, often wriggling into loose scree and boulder piles on the cave floor.

Brown Long Eared bats have the coolest requirements when it comes to winter roosts, often choosing to sleep near the entrance of the cave. When asleep they fold their enormous ears backwards under their wings.

LAW AND PROTECTION

The most important thing to know is that bats must not be disturbed; in fact, under the 1985 Wildlife Order, they are protected by law. It is an offence to kill, disturb, handle or offer for sale any bat, alive or dead. Similarly, it is an offence to damage, destroy or block access to any place that they use as shelter. There is a licensing system which permits trained individuals to handle and carry out research on bats.

During hibernation, their body temperature drops to that of their surroundings and their heart, respiration and blood circulation all slow down proportionally. They may wake up a couple of times over the winter, shivering as they restart their metabolism. A lot of energy is required to bring their temperature up to the 42 degrees centigrade necessary to enable flight. Disturbance of hibernating bats can be catastrophic. Shining a light or making a noise may be enough to wake them. The vital energy reserves lost by the disturbance can tip the fine balance between living through to springtime or dying.

Occasionally, metal grilles are placed across entrances to sites where there is a cave roost, to ensure that there is no disturbance. They are designed carefully to avoid altering the climate of the roost or interfering with the flying bats.

Subterranean water also contains temporary cave dwellers. Although they may not have entered by choice, fish are often seen in caves. They are washed along in the river as it flows underground and re-emerge in daylight, seemingly no worse for the experience. Frogs too, are often seen deep underground. They can survive as long as they have a food supply.

When Black Burn Cave in Antrim was first entered, two white trout were found swimming lazily in an isolated pool in the white limestone. Their lack of reaction to the caver's lights fuelled a theory of blindness and the newspapers picked up on the story and expanded the tale into one of "blind white cave fish discovered in County Antrim". Discussion with biologists revealed that their lack of colour was a temporary loss of pigmentation, a normal reaction to remaining in total darkness for some time. Their apparent lack of interest in the caver's light was probably a result of starvation. When washed out to the surface in the next heavy rainfall they would recover completely and their pigmentation would return.

There are other, less frequently seen, cave visitors. In one river cave in Fermanagh, a caver had the surprise of his life when, about a hundred metres from the entrance, he disturbed an otter feeding on fish in complete darkness at the water's edge.

■ OUR PERMANENT CAVE-DWELLERS

Finally, we come to perhaps the most fascinating ones, the ones we know least about: the troglobites. All have evolved from surface-dwelling ancestors which, at some point in the dim and distant past, have moved into caves.

After ages of isolation underground, the troglobites adapt to a life in total darkness. Their eyes disappear - sight is no longer of any use. There is no need for protection from the sun or for camouflage from enemies, so pigmentation vanishes. But to survive they must develop highly sensitive sensory organs to detect food and predators.

Here in Ireland, we may not have the amazing salamanders with their pure white bodies, long legs and eyeless faces but we do have troglobites of our own. They are small and almost invisible to the human eye, but they are troglobites nonetheless. They are insects of the sort called springtails, some species of which have become cave-dwellers.

The Caves and Limestone Scenery of the North of Ireland

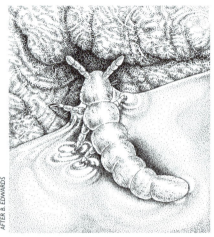

Fig. 99 *An artist's impression of one of the Springtails (Collembola) which inhabit our caves. They measure no more than 1mm and are barely visible to the naked eye.*

■ OUR OWN TROGLOBITES

At present, there are three known troglobites in the caves of the north of Ireland, out of the eight in Ireland as a whole. They are all arthropods called springtails (*Collembola*). These minute creatures with hard, jointed external skeletons live, floating on the surface of pools, crawling over mud banks or in among piles of bat droppings. They feed on moulds, bacteria and organic detritus so they are ideally suited to cave life. They, in turn, are preyed on by flat worms in the pools and by spiders.

Of the three, *Onychiurus schoetti* is the most interesting and has only been found in two Fermanagh caves so far. It has a slender, white, rod-shaped body with short club-shaped antennae and sharp claws. The claws help the springtail to clamber up the calcite dams of gour pools in search of new feeding grounds. It is found in caves in England and Europe, but only on the surface in Norway, in leaf mould. This is seen as evidence that it is a species which prefers a cooler climate so, as the last glaciers retreated north and the climate warmed, it moved underground to find conditions more to its liking.

Troglobites can never return to a life above ground; their bodies have undergone permanent changes. Cavernicoles, however, have a return ticket if necessary.

There are a number of cavernicole populations in our caves that include species of beetle, spiders and fungus gnat. In Waitomo Cave in New Zealand, the larva of one particular species of fungus gnat is the 'glow worm', a major tourist attraction.

In our caves, the most commonly noticed cavernicoles are white flatworms and fresh water shrimps. Because they are commonly found in surface streams, it is often difficult to say if a population is actually maintained in a cave. In a pool in Pollaraftra, Co Fermanagh, high above the river, shrimps spin around among pebbles and the soft mud bottom, (Fig. 77). As this area only floods occasionally, this population is thought to be permanently cave dwelling.

Flatworm populations have only been recorded in the regularly flooding passages in Pollasumera and Black Burn, so it is difficult to tell if these are true cave populations. They are scavengers, cruising just beneath the surface film of the water and pouncing on any food, living or dead. In Black Burn Cave, they are difficult to see, as their loss of colouring means that they are now actually camouflaged against the pure white of the Antrim limestone.

As well as the 'big' full time residents, there are also a host of the simplest forms of life; the bacteria, algae and fungi. They don't readily come to mind because there are not easy to see but they play an important role in the story. They use organic material and minerals to produce the energy they need to eke out a living on cave walls and mud banks. They are nature's smallest recyclers. Right at the bottom of the food chain, they are important food sources for others.

Perhaps the most conspicuous example of bacteria and algae in action is found in the pasty, white, calcareous substance known as moonmilk, where they seem to influence the crystal growth of the carbonate minerals. Moonmilk is, however a bit of a mystery. The way it is formed is not fully understood and it is difficult to identify, as it looks very similar to soft calcite deposits. We cannot even say for certain to what extent it occurs in the caves of the north of Ireland. The only way to find out would be to take samples and study each mass in detail.

■ ALL THAT GLISTENS IS NOT GOLD

People entering a cave for the first time often notice sparkling gold or silver patches on cave walls. They might be disappointed to learn that these patches are not hidden treasure but what is sometimes called 'wall fungus'. The colonies measure 1-3cm across and are white, sometimes with a yellow centre. Their enticing, metallic appearance arises because the droplets of condensation hanging on them pick up the torch light making them glisten and twinkle. They occur most frequently near entrances and are largely responsible for the characteristic earthy smell associated with caves. Like moonmilk, they are still a bit of an enigma but they are thought to be a fungal/bacterial combination. Outside caves, this association is known for its ability to produce antibiotics.

Fig.100 Fungi grows on flood debris in complete darkness far underground.

Cave biology is in its infancy both here and in caves world-wide. There is much to learn about these isolated, highly specialised populations. One thing is certain, however: cave dwelling communities are extremely delicate and susceptible to disturbance. Their presence is of great importance if the planet's biodiversity, its variety of life, is to be maintained.

Fig. 101 'Wall fungus' shines on the roof of a Fermanagh cave.

The Caves and Limestone Scenery of the North of Ireland

Beneath our Feet

The Caves and Limestone Scenery of the North of Ireland

People and limestone

7

PEOPLE AND LIMESTONE

Our limestone landscape is not just home to lime-loving plants and cave-adapted bugs and beasts - it also supports a thriving human population. Over time, people have shaped their way of life to suit its thin soils and scarcity of water. At the same time, they have made good use of its unique properties and in so doing, have shaped the very landscape itself. Glance around the limestone countryside today and you will see farmhouses and barns, quarries and factories. Look harder and longer and you will see evidence of older habitation, proof that the limestone has attracted and supported communities for, not hundreds, but thousands of years.

■ FROM STONE AGE TIMES

After the last ice sheets finally retreated, plants and animal communities transformed the bleak landscape. People completed the picture when they travelled from the west coast of Britain to Ulster, around 9,000 years ago, at the start of the Mesolithic or Middle Stone Age. They settled first in the east of the country, particularly attracted to County Antrim and its supplies of flint. The flint, seen as protruding brown lumps in the Ulster White Limestone (Fig.45), was at that time the most effective material available for tool-making and found only in this part of Ireland. It was used to make all sorts of implements: axes, picks, scrapers, arrow heads and harpoon barbs. Evidence of flint-working and flint tool manufacture along the Antrim coast represents one of the earliest known examples of natural resource exploitation in Ireland.

These first Mesolithic people were nomadic, their homes no more than makeshift shelters. They roamed the dense forests, fishing, hunting and gathering, slowly moving westwards across the country. Apart from their tools, they left few other tangible remains. Seven mudstone axes and a double-pointed pick were discovered in Cushrush Island, beneath Hanging Rock in Fermanagh. These were dated to the late Mesolithic, 7000 years before the present. They serve as convincing evidence that the limestone hills of Fermanagh were inhabited from that time. Chert, the silica deposit common in some parts of Fermanagh limestone, was also used to make tools.

Around 6,000 years ago, a new wave of incomers arrived, bringing with them a radically different way of life. They were the first farmers of the Neolithic or New Stone Age. They brought sophisticated implements and knowledge of animal husbandry and crop tillage. They ring-barked and burnt the trees to make way for their agriculture. When the soil declined in fertility, they moved on and did the same thing again to another patch of forest.

An enduring legacy of these first settlers is their giant stone burial monuments or megalithic tombs still standing, strong and prominent, in the landscape today. In spite of the numerous caves which could have served as ready-made graves, it seems that the farmers chose instead to bury their dead in these dramatic structures. They were often sited on hilltops or terraces, commanding splendid views to the lowlands. Huge stones were lifted

The Caves and Limestone Scenery of the North of Ireland

Fig. 102 Greenan wedge tomb situated on a limestone terrace looking east to Knockninny and Upper Lough Erne.

Fig. 103 A pine stump preserved beneath a blanket of peat provides clear evidence that forests once covered the land.

and transported by ingenious means and considerable muscle-power to these special locations. In the limestone areas of Fermanagh, whenever possible, the builders chose as their building material the sandstone erratics which were lying ready and available for use, a much easier option than quarrying the limestone bedrock.

Over time, the influence of these and succeeding farmers on the land was profound and lasting. Indeed, as we saw in the previous chapter, our modern landscape is, to a considerable degree, the legacy of these prehistoric inhabitants. Forest clearance, coupled with the onset of a cooler, wetter, climate, led to the eventual accumulation of peat and the creation of great expanses of bog. Even on the well-drained limestone, peat smothered the rock. Up on East Cuilcagh, where dense woodland once flourished, the limestone is almost completely hidden from view. Here, a megalithic tomb is just visible, emerging from the bog. It hints at the presence of an ancient, farmed landscape, preserved beneath the blanket of bog, similar perhaps to the famous prehistoric landscape at the Ceide Fields in County Mayo.

Confirmation of the cutting and burning of the forest can be seen in sediment banks in the far reaches of the Marble Arch Caves (Fig. 40). The undisturbed deposits contain a distinctive charcoal layer, which has been related to that major event in the evolution of the landscape.

Cave sediments also provide a valuable store of pollen. Pollen grains are exceptionally resistant to decay. Furthermore, the pollen grains of most types of plants or trees are different from each other; their protective outside walls are textured and etched into intricate patterns, a bit like human fingerprints. Pollen can therefore be analysed and so provide us with information about what the landscape was like at any one time. If, for example, analysis of a sample of sediment shows a marked drop in tree pollen coupled with the appearance of grass pollen, it suggests to us that Neolithic farmers had arrived in the area and were clearing the forest to make way for their crops.

In County Antrim, there is evidence that the fertile soils of the white limestone have supported generation after generation of farming communities. For example, Linford, near Ballygalley, is an area of grazed limestone which has remained unchanged and undisturbed for centuries, in such a way that Bronze and Iron Age earthworks, clearly visible today, are superimposed on sites dating back to Stone Age times.

Similarly, in the aptly named townland of Goodland near Ballycastle, there is remarkable evidence of over 100 hut sites and associated road and field patterns preserved as low grass-covered banks on the limestone. The exact purpose of the settlement is unclear but again it suggests that this fertile site sustained thriving farming communities, perhaps for booleying, the ancient farming practice of summer grazing of stock on higher, fertile pastures. Booley huts were the small shelters built by the farmers for use during the summer months. Again there is evidence that these remains overlie an older farmed landscape. Archaeological excavations revealed a Neolithic settlement consisting of

The Caves and Limestone Scenery of the North of Ireland

ditches and pits filled with charcoal, broken pottery and flints. A collection of decorated, globular-shaped bowls were also discovered; such bowls found elsewhere in Ulster are also called Goodland Bowls.

County Antrim has numerous similar sites of archaeological interest on the outcrops of limestone, testament perhaps to the enduring attraction and value of the rock and its sweet grassland.

■ CAVE ARCHAEOLOGY

We have no evidence to suggest that our early ancestors actually lived underground in this part of Ireland; perhaps the caves were too draughty or too wet but there is certainly proof that they lived on the land around them. In 1972, a caver noticed a collection of bone material at the bottom of a small pothole called Pollthanacarra, on the Marlbank in Fermanagh. The Ulster Museum carried out an excavation and removed large quantities of bones which, when identified and carbon dated, were discovered to be the remains of three humans as well as red deer, wolf, boar, cattle and hare (dated at 4,500 years before the present, the Neolithic/Early Bronze Age). There is no way of knowing how the three humans met their death, whether they fell or were thrown down and no artefacts were found with them. The study did, however, reveal that they were sturdy and well nourished. The pothole, at that time, would have been concealed in dense undergrowth, so it can be assumed that it was a trap for the unwary. This theory can be borne out by the fact that the majority of the bones excavated were of young animals, victims perhaps of their own inexperience.

Although it is unlikely that our caves were used as permanent dwelling places there is evidence to suggest that the entrances of some may have provided temporary places of repose or retreat from the wind and rain, particularly those which commanded views over the surrounding countryside. For example, a small, relict cave on the slopes of Knockninny Hill beside Upper Lough Erne, is said to have been the hermitage retreat of St. Ninnidh, one of St Patrick's diciples. Knockninny is named after St Ninnidh, who, during the sixth century AD, came each year to the cave to fast for the 40 days of Lent. The hill also boasts a spring rising, known as St Ninnidh's Holy Well. Excavations in the sediment floor of the cave was undertaken by a local amateur collector-cum-archaeologist named Plunkett in 1875. He uncovered fragments of pottery and an urn containing cremated remains dating to the Bronze Age.

Plunkett also excavated the earth floors at the entrances to other caves in Fermanagh to find layers of charcoal, broken bone and fragments of pottery. One such small, rock shelter on the top of Knockmore, named the Lettered Cave, has scribings etched into the wall, now sadly overlain with more modern graffiti.

In Antrim, the caves in the sheltered southern shore of Rathlin Island have long been used for shelter and storage. During the summer, fishermen from the mainland slept in them by night and fished during the day. The caves were owned by different island families and were used to store their lobster pots.

Fig. 104 The interlaced crosses and knots may date to prehistoric times although it is more likely that they are early Christian in origin, possibly dating from a time when the cave was used as a secret place of worship.

The Caves and Limestone Scenery of the North of Ireland

Fig. 105 A cup and ring marked stone lies on the limestone near Boho. The purpose of these carvings is still a mystery.

Fig. 106 Holywell, near Belcoo. This well never dries up even in the hottest summer. In pre-Christian times it was a site for pagan ritual. The Celtic sun god, Lugh, was honoured here every year, during the festival of Lughnasa, to give thanks for the first fruits of autumn. Its significance may be indicated by the concentration of prehistoric remains centred around it. Later, it became a place of Christian pilgrimage, reputed to have been visited by St Patrick. Today, at the beginning of August each year, barefooted pilgrims maintain the tradition.

With the gradual transition to the Bronze, then Iron, Age, different monuments appeared in the landscape, particularly stone circles, standing stones and rock art. There are many theories as to what function they served; some were ritual sites, some used for burial, some perhaps for navigation or understanding the stars. Whatever their original purpose, they now make distinctive landmarks.

One particularly fine example of rock art is the Reyfad stone near Boho; these sandstone erratics resting on the limestone are carved with hollows and rings known as cup and ring marks.

Caves in the north of Ireland may not have been used as permanent homes or as burial sites as they are in other parts of the world, but they most certainly had great mystical significance in Pre-Christian times. The sight of rivers bubbling out of the ground or plunging into the black depths of the underworld must surely have fired fertile imaginations. Resurging streams and wells were often sites for pagan ritual around which places of worship and habitation developed. The waters of many were believed to have curative properties. On the south facing slope of Belmore Mountain, there are a number of these sites, including a jaundice well, a scurvy well and an ague well. The wells around Swanlinbar on the Fermanagh / Cavan border were so famous during the 19th century that the town built up a reputation as a spa town, attracting people from all over Ireland to 'take the water'. Many people today still swear by the beneficial effects of spa water, in spite of its often evil taste. Perhaps they rely on the premise that, if it tastes disgusting, it must be good for you!

■ KILTIERNEY

Kiltierney Deer Park in County Fermanagh is a remarkable place. Within this small, isolated patch of rolling limestone farmland with its cave and river sink and rising, is an amazing concentration of archaeological sites. Clearly visible are a Neolithic passage grave, barrows, an Iron Age burial site (the only proven one in Ulster), raths, two holy wells and a church (part of a Cistercian grange): sites which span the history of civilisation, Christian sites superimposed on pagan sites. An indication, in a nutshell, of the immense mystical and spiritual significance of the limestone landscape from earliest times.

Fig. 107 A limestone arch straddles the County river and links the counties of Leitrim and Fermanagh. Beside is a monastic site dating back to early Christian times. It is said that St Patrick used this natural bridge on his journey from Ulster to Connacht.

The Caves and Limestone Scenery of the North of Ireland

In early Christian times, people lived in defensive sites called raths or forts. These were circular enclosures, consisting of a bank and outer ditch, built to house the extended family of wealthy farmers and provide defence from wolves and hostile neighbours. In limestone areas and other rocky sites where there was insufficient earth to dig a ditch, the circular walls were built of stones and are called cashels. Occasionally, raths had underground storage chambers, known as souterrains, excavated in the ground within the walls of the enclosure, often doubling as refuges from raiders. They were unusual in limestone areas; the hard rock would have been difficult to excavate. However, there is one of note near Boho, which cleverly incorporated a natural cave passage: an ideal cool, dark store for perishable food.

An unusual use of caves occurred during the Penal Years, in the 17th and 18th centuries, when they were used as mass sites, places where people could come to worship in secret. Some caves still retain names which date from this use, like Pollanaffrin, from the Irish '*poll an Aifrinn*' meaning cave of the mass rock.

■ FOLKLORE

There is a wealth of folklore associated with caves which has been passed from generation to generation and richly embellished over time.

A common and recurring theme is the one that tells of a musician, usually a piper or a fiddle player, who ventures underground and is never seen again. Of course his ghostly playing can still be heard to this day! Fiddler's Cave in Kiltierney is the home of one such ill-fated adventurer.

Another is the tale which concerns the unfortunate dog which disappeared into a cave on one side of a mountain and emerged from underground at the other side many days later, its fur singed from the flames of the underworld! Belmore Mountain boasts versions of this story.

Potholes are generally considered bottomless, even when they have been fully explored, surveyed and their depth measured; perhaps a cautionary tale told to overly-inquisitive children.

One story about the deepest pothole in Ireland is, however, fact rather than fiction. Noon's Hole in Fermanagh is named after Dominic Noone who, during the 1820s, informed on an outlawed society called the Ribbon Men. He was tricked into attending a wedding party, then kidnapped and taken to the Barrs of Boho and thrown down the nearby pothole. Some stories say he fell the 30 metres to the first ledge and when his cries for help were heard, someone was lowered on a rope, not to rescue him, but to kick him the remaining 70 metres to the bottom. Other versions say that his body was recovered from the first ledge by a quarry man, who was lowered on a rope to retrieve him. The gory details are contained in a book called 'The Ribbon Man Informer", written in 1874.

■ EXPLOITING THE RESOURCE

Earlier chapters have left us in no doubt that limestone is a strong rock, resisting the elements and forming dramatic cliffs and prominent knolls. It follows therefore that, once people had acquired the tools and techniques to break it, the stone would be in demand as a building material and also for its lime. Limestone, when subjected to intense heat, is reduced to calcium oxide (CaO), sometimes called burnt or quicklime, that has had a great range of uses since Roman times: for spreading on land as a kind of fertiliser, for use in the bleaching process in linen making, for white-washing walls and for making building and plastering mortar.

At the height of its demand during the 19th century, every community would have had access to a lime kiln. The fact that over 15,000 lime kilns are marked on O.S. (N.I.) maps gives some indication of just how important burnt lime was. In areas like County Down where there is no local limestone, kilns were built beside the sea so that limestone could be shipped round the coast to them.

Fig. 108 Gortalughany lime kiln, perfectly sited on the road beside the quarry and with a ready supply of turf to fire it nearby. Kilns were often built into the bank or level with the road to make them easier to load.

To make the lime, the rock was first broken into small lumps and layered in the thick-walled kiln with wood and turf and then lit. It would be kept smouldering overnight, with more turf and rock fed in from the top. The commercially run kilns might be kept burning for months on end without cooling. The burnt lime was recovered from the small opening at the bottom. In Antrim, the lime kilns were often fired using the local supplies of coal and lignite, as well as turf.

Production of lime declined during the 20th century with the development of powerful grinding machinery. Cement, crushed lime and artificial fertilisers slowly replaced lime, and kilns ceased to function. However, they can still be seen today in various stages of decay; some are small, just big enough to produce lime for a few families, while others are huge, like the one overlooking the Carrick-a-rede car park in County Antrim, which was in operation up until the 1970s.

In spite of the fact that cement is a superbly strong, quick setting building material, lime has some qualities which cement cannot match. Lime mortar lets buildings 'breathe', so preventing the build up of moisture. It also flexes without cracking and it contains fewer harmful salts. Because of these qualities, lime, rather than cement, should be used in the repair of historic buildings.

With a revival of interest in lime, a few new kilns have been opened, like the one at Narrow Water near Newry which brings its limestone from Antrim, not only to supply lime for the upkeep of historic masonry, but also for use in modern construction.

Limestone quarries of various sizes are scattered across the land. Some are close to lime kilns and long abandoned, no more than a few blocks removed from a small outcrop, while others are huge commercial operations, devouring whole hillsides and leaving

The Caves and Limestone Scenery of the North of Ireland

permanent scars on the landscape.

The Antrim coast is dotted with limestone quarries particularly around Glenarm and Carnlough where, in the mid 19th century, the Londonderry family constructed the present day harbour and linked it to the quarries with a cable-operated railway. From the harbour, limestone was exported to Britain where it was used as a flux to draw off impurities during iron smelting.

Nowadays, limestone is in great demand as road stone, and as the principal ingredient of Portland cement, so named because of its supposed appearance and similarity to Portland limestone from the south of England. Portland cement is made from limestone, which is pulverised and fired at high temperatures with compounds such as silica to give it added strength and gypsum to aid setting. Concrete, one of the modern world's chief building materials, is made by mixing cement with an aggregate such as sand.

Pulverised limestone rather than burnt lime is now spread on arable land. The powder is important in the cultivation of cereals as it is easily absorbed and helps neutralise acitity.

Belmore Mountain in Fermanagh is valued for its concentration of high quality dolomite; the largest economic deposit of this rock in the north of Ireland. Dolomitic limestone with its high magnesium content has an agricultural value as it is used to treat a magnesium deficiency condition in cattle, known locally as the staggers.

The Eglinton Limestone Company in County Antrim draws raw material from the Demesne Quarry to its roadside Whiting Mill in Glenarm. The rock is ground to a fine white powder for use as an industrial filler in products ranging from cosmetics to paint to toothpaste. A far cry from its origins on the sea bed!

■ TODAY AND YESTERDAY

Changes in the present day limestone countryside echo the changes taking place in rural areas elsewhere. Many of the remote farmhouses are derelict, their occupants driven from the land by the famine, or abandoned in favour of an easier life on the lowlands. Stone outhouses have been replaced with large steel sheds. Small fields have been enlarged to make access easier for farm machinery. Many of the stone walls have tumbled down or been replaced with fences; hedges and trees have been cut down.

Now, this landscape, once seen as remote and less favoured, is enjoying a surge of interest as its special qualities are realised and appreciated. It is facing new challenges created by the demands of tourism and recreation. However, tourism is not a new phenomenon. From the time people had money and time to travel, this landscape has attracted interest. From the early 1700s, the Earls of Enniskillen, who resided in Florence Court House, invited eminent scientists of the day to Fermanagh and conducted field trips through the surrounding limestone scenery and also along the Antrim coast. In 1796, the French

The Caves and Limestone Scenery of the North of Ireland

traveller and writer, De Latocnaye, published his book "A Walk through Ireland" in which he describes a visit to the Marble Arch. Even then, there were local guides keen to give candlelit tours of the caves. Sites like Coolarkan Cave near Boho were popular picnic spots for the gentry of the day where they ventured a short distance underground with flaming torches to marvel at the underground world. Many years later, in 1985, the Marble Arch Cave was to be developed as a show cave and is a major present-day tourist attraction bringing thousands of people to the limestone of Fermanagh.

Beneath our Feet

The Caves and Limestone Scenery of the North of Ireland

Cave exploration - the sporting science

CAVE EXPLORATION - THE SPORTING SCIENCE

The earth has been photographed from space, its surface has been mapped in detail, its highest mountains have been scaled and its remotest deserts scrutinised. There are even footprints on the moon. Where else on this crowded planet, apart from the deep oceans, are there natural places remaining which offer the potential for true, original exploration and study, other than caves?

We will know when the last mountain has been climbed but we can never say, for certain, that the last cave has been found.

Cave exploration is not a recent phenomenon. 2,200 years ago, a book entitled "The Mountain Scripture" was written in China describing caves and cave formation. At a time when the inhabitants of the limestone landscapes of Ireland were, most likely, regarding caves with superstition and fear, the Chinese were going underground and describing how a drop of water falls from the tip of a stalactite.

Today, this study is called **speleology**, coming from the Greek word for cave, spelaion. So, the exploration of caves is more than just a physical challenge, it is a science. And, as we have seen throughout this book, it encompasses a wide range of 'ologies'; biospeleology, archaeology, palaeontology, hydrology, geology and the list goes on. But there is one thing they all have in common; their study requires the ability to enter and explore caves safely. Not only should safety be considered paramount for the explorer, it is also paramount for the cave. There is little use in discovering and studying a wonderful new environment, if it is damaged or destroyed in the process.

A HISTORY OF EXPLORATION

■ PIONEERING DAYS IN FERMANAGH

Venturing underground cannot be treated lightly. Caves are wild places. They have been carved by flowing water. When it rains, the caves react. They fill, often to the brim. And although carved in solid rock, we have seen that they have areas of collapse - often huge boulders the size of buses are precariously balanced one on top of another. They have unexpected hazards: pits in the floor, deep lakes, fast flowing rivers. There is no light whatsoever. If one's light source fails, there is no possibility of finding the way out without help. Here in Ireland, caves are chilly by human standards, cold enough for hypothermia to be a very real and constant threat.

It is not surprising therefore that, when all the hazards are considered, the caves of Fermanagh were not systematically explored until the end of the 19th century, although they certainly had been looked into for a short distance by inquisitive locals and visitors prior to this. Local people tell of how they made rafts, stuck candles on them and set them adrift down the cave rivers to illuminate the passage beyond the point they were willing to delve.

But the first person to actually study them scientifically was a Frenchman named Edouard Martel, a lawyer by profession but a fanatical speleologist. His consuming passion for caves took him all over Europe and in 1895, he came to Ireland to explore caves, starting with the Marble Arch. He was accompanied by Lyster Jameson, a student of entomology from Dublin who had been given a grant by the Royal Irish Academy (prompted by Martel's visit) to investigate cave fauna in Ireland. By the flickering light of candles and miner's lamps, they manoeuvred their folding canvas boat along the peaty brown waters of the Cladagh River. They were not dare-devils but scientists, carrying out systematic study; they took measurements and collected data e.g. air temperature and humidity. They used magnesium flares to illuminate the large chambers and, using this information, drew underground maps and wrote detailed descriptions.

As well as simply walking or boating into caves, Martel was also prepared to take on the 'bottomless' by being lowered down the first 20 metres of the Noon's Hole entrance shaft and while doing so, looking around and making some accurate observations as to how it was formed. Imagine trusting your life to a piece of hemp rope tied round your waist, as you are lowered down, inch by inch into an abyss, a waterfall tumbling on your head threatening, at any minute, to extinguish the feeble flame of the candle stuck in the brim of your hat. The rope is sending you spinning round and round in space. Imagine looking up and seeing that daylight is only a speck of light far above. Imagine standing cold and shivering on a ledge, shouting at the top of your voice, in an effort to make yourself heard over the echoing din of the water, as you attempt to tell your friends above that you're ready to start the long, slow haul out.

Later, Martel also visited the Antrim coast and explored and described the Larrybane Caves. By the time of his death in 1938, he had investigated nearly 1,500 caves and his technical innovations had become standard equipment for cavers, inspiring a new generation who would continue the work of this 'infant' science in Europe. It seems appropriate therefore that he is known as 'the founder of modern speleology'. In his book, 'Irlande et Cavernes Anglaises', Martel described his explorations. His writings and lectures inspired an English group of adventurers called the Yorkshire Ramblers to visit Fermanagh and continue the exploration.

From 1907 to 1947, the Yorkshire Ramblers conducted 10 expeditions to the Fermanagh area, constantly furthering the knowledge of the caves. Robert Lloyd Praeger, Ireland's great naturalist, accompanied them on their 1907 exploration of the Marble Arch Caves. Afterwards, he wrote of his exploits saying, "if you want an unusual experience, try cave-lake swimming, the water is cold and as black as ink".

As well as continuing the exploration of Marble Arch, the Yorkshire Ramblers also surveyed many other systems in the Marlbank area of Fermanagh and also around Boho and Derrygonnelly, such as Pollaraftra. With heavy ropes, wood-runged ladders, tweed jackets, and hobnailed boots, they made some impressive underground sorties. They were tough characters, able to put up with considerable discomfort in their quest for discovery. In 1912, two intrepid members of their group, Wingfield and Baker, made the first descent to the bottom of the entrance shafts of Noon's Hole.

Fig. 109 Baker descends Noon's hole on a rope ladder. Although Baker and Wingfield reached the bottom of the shaft at -100m they didn't actually find the way on as they had descended the wet shaft, missing the parallel dry shafts. It was not until 1970, that Leeds University Speleological Society descended the dry shaft and thereby found the key to the cave system.

These early exploits required teamwork, planning and determination. However, these pioneers considered the rewards were worth all the pain. Indeed, the rewards were the same for them as they are for today's cave explorers, the chance to shine the first light on a hidden corner of the world.

In 1939, the Yorkshire Ramblers also descended the entrance shafts of Reyfad, now known as one of the longest and deepest cave systems in Ireland. Each year that they returned, their equipment became more sophisticated. They used carbide bicycle lamps mounted onto brackets on climbing helmets. These lamps are still popular with cavers today as they shed a diffuse, bright light which perfectly illuminates large chambers.

The Caves and Limestone Scenery of the North of Ireland

The visits by the Yorkshire Ramblers soon inspired other British clubs to come across and also encouraged local folk to get involved. When one group reached their limit in exploring a cave, another would take over, perhaps finding the way through a complex boulder pile, or descending a pitch or climbing a shaft. Lightweight wire ladders, used in conjunction with nylon mountaineering ropes, were an important development, as they proved much less cumbersome than the old rope ladders.

The incentive was always to find more cave passage, to fill in the missing pieces of the jigsaw between the sink and resurgence. Caves systems lengthened, deepened and became more complex, so demanding more commitment and skill from the explorers.

Fig. 110 Flexible wire caving ladder and lifeline in use on a pitch.

■ A CAVER BY ANY OTHER NAME

The question is often asked, "what is the difference between a caver and a potholer?" The answer is simple. Caving involves exploring horizontal passages: walking, swimming, squeezing between boulders and crawling flat-out along bedding planes, whereas potholing is the exploration of vertical fissures. This involves the use of technical equipment: abseiling down ropes to get in, climbing up the ropes to get out - skills not for the foolhardy or ill trained. The knowledge and ability to extricate yourself from difficulty is essential. Many of the shafts of East Cuilcagh are purely potholing trips as there is no open, horizontal cave at the bottom. Many other trips involve both caving and potholing. Noon's, for example, starts as a potholing trip to negotiate the entrance shaft, then continues as a caving trip to follow the water along its horizontal journey.

Spelunker is the American term for caver, while the term used internationally for a caver and cave scientist is speleo. They all mean the same, someone who ventures underground.

Cave diving involves the use of scuba equipment to negotiate sumps, water filled passages sometimes called siphons, in the hope of finding dry passages beyond. This method of exploration requires the use of carefully adapted diving equipment, and the laying of route-finding lines. It needs rigorous training, meticulous planning and leaves no leeway for error. The brown, peaty waters of Irish caves complicate matters further by reducing visibility to almost zero. The murky water often defeats the most powerful of lights making it almost impossible for cave divers to see their way on.

The Caves and Limestone Scenery of the North of Ireland

■ THE PACE OF EXPLORATION QUICKENS

By the mid 1960's, a number of Irish caving clubs had been formed and exploration went on apace. Coal miner's electric cap lamps, with a rechargeable battery worn on the waist, provided waterproof, bright and reliable lighting. Neoprene rubber diving suits were now worn in wet caves giving much more insulation against the cold water than ordinary clothes. They allowed cavers to stay warm for longer, and therefore underground longer.

Equipment and clothing were also being developed specifically for cave exploration. Reyfad Cave was now so extensive that, to carry out useful exploration, cavers were camping underground. Distances which, on the surface, took only minutes to stroll across, took many hours to traverse underground, such is the depth and complexity of the cave.

A group of local cavers calling themselves the Reyfad Group, made major discoveries during this time, not only in Reyfad but also the Cascade System on the Marlbank, Noon's and numerous others. Everything they explored was methodically surveyed, the geology studied and detailed maps produced. Eventually, in 1974, a guide book of the caves of Fermanagh and Cavan was produced. One of their most significant and later discoveries was Shannon Cave. A stream, flowing on the slopes of west Cuilcagh in Fermanagh, was followed underground as it flowed into County Cavan; caves know no political boundaries! The underground waters eventually emerge from Shannon Pot, the source of the mighty Shannon River.

Fig. 111 A cave diver prepares to explore an underwater passage.

The Caves and Limestone Scenery of the North of Ireland

Fig. 112 *Surveying caves can be quite a challenge. Here, the caver has to lie flat out in the stream bed to map this small passage.*

■ CAVE SURVEYING

It is impossible to have an idea of what a cave looks like until it has been surveyed and a map drawn of it. This involves taking a number of measurements along each 'leg' of the survey. A leg is the line of sight between two points called 'stations'. When the cave takes a change in direction, a new leg is started. Three measurements are taken from each station:

1. The distance between stations, using a waterproof tape measure.
2. The direction in which the leg is trending, using a magnetic sighting compass.
3. The dip or rise of the passage, using a clinometer.

An accurate line survey can then be drawn, showing length, direction and elevation of the cave. Detail is added: heights, widths and features e.g. rivers, stream inlets, stalactites and boulder chokes. All this data is noted down in a waterproof notebook.

The measurements are then adjusted by trigonometry to 'flatten' the passage, allowing for all the dips and rises. The result is a detailed, accurate cave plan. It can clearly illustrate the geological controls which have determined the character of the cave e.g. maze-like if the limestone is heavily jointed, a straight passage if the cave follows a major fault etc.

The survey can be superimposed on a surface map so that it is possible to say where a cave is running in relation to the landscape, as shown in Figures 60 and 74.

The most significant development in cave equipment took place in the 1970s, with the development of Single Rope Technique (SRT). It involved the modification of rope climbing clamps originally developed for mountaineering. These, along with the development of low stretch caving rope, allowed a caver, having abseiled in, to literally 'walk up' the rope to get out, without the need for an additional lifeline.

The development of SRT, coupled with the introduction of warm, synthetic fleece undersuits and waterproof oversuits means that, nowadays, exploration is much more efficient and comfortable. Small groups of people can move quickly and safely, exploring further.

Fig. 113 *With developments in caving equipment, cavers are able to explore remote, previously inaccessible passages. Here a potholer climbs safely around a shaft on ropes.*

The Caves and Limestone Scenery of the North of Ireland

■ SPELEOLOGICAL DISCOVERIES OF THE 1980S

Black Burn Cave in County Antrim was not discovered until 1984. A group of canoeists were out walking, looking for new rivers to negotiate, and instead they discovered a stream disappearing into a cave! They passed the news of their find on to cavers. The exploration of Black Burn to its far reaches was only possible by cave diving, as it contains a number of sumps.

The golden years of easy discovery of caves in Ireland are finished but there is still much information to gather from the known caves. Today, scientific research reveals more and more about them. Detailed geological study, dye tracing, water chemistry, along with observation and intelligent guesswork, further the knowledge and also help in the search for new passages, but cavers must work hard for their chance of original exploration.

■ PHOTOGRAPHING DARKNESS

Often, paradoxically, it is not until cavers come out of a cave and see the photographs of it, that they can really appreciate the beauty or grandeur of where they have been. But how is it possible to photograph darkness? Good cave photography is certainly an art and one that can take years of learning, along with rolls of wasted film, written-off cameras and some very cold and fed-up assistants. Generally, caves absorb a lot of light. Powerful, synchronised flash bulbs or guns are needed to efficiently illuminate the scene. When results are successful, all the effort is worthwhile.

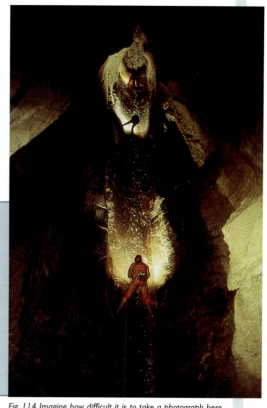

Fig. 114 *Imagine how difficult it is to take a photograph here. The caver, communicating with the photographer, has triggered a hand-held flash gun three times to light this shaft. Between each flash she has ascended the rope in the dark while the camera shutter has been kept open.*

■ OUR CAVES IN A WORLD CONTEXT

This list is already out of date! As soon as it was compiled, a cave passage somewhere around the world was being discovered. Old limits are constantly being pushed. There is always a thirst for discovery of new sites and new records in depth and length.

■ LONGEST AND DEEPEST

Ireland's Longest:	Pollnagollum-Poul Elva System Co. Clare-14.5 kilometres
Fermanagh's Longest:	Reyfad-Pollnacrom-Tullybrack-6.7 kilometres
Ireland's Deepest Pothole:	Noon's, Co. Fermanagh-92 metres
Fermanagh's Deepest System:	Reyfad-Pollnacrom-Tullybrack-193 metres
Longest in the world:	Mammoth Cave System, Kentucky, USA-530 kilometres
Deepest in the world:	Voranja, Georgia-1710 metres

The Caves and Limestone Scenery of the North of Ireland

> ### ■ ORIGINAL CAVE EXPLORATION IN FERMANAGH
>
> *Extracts from a caver's diary.*
>
> *I lay flat out, peering into the dark space. The boulder we had just pulled aside had revealed a tantalising black hole beyond. I tried to squeeze my head into the rift to see if there was a way on. It was too narrow. I couldn't shine my light in and look at the same time. I pushed another small boulder aside, the gap was wider and now I could just about see that my light was shining into darkness. I was sure I could hear the deep enticing music of flowing water far ahead, but maybe my over active imagination was playing tricks on me. One thing I was sure of, there was a howling draught blowing in my face. It made the flame of my carbide light flicker and hiss. If anything hinted of passage ahead, it was this wind. My companions waiting behind me were getting bored and cold. "Can you see anything, is there a way on?" I squeezed in, then backed out to make sure I wouldn't get stuck. I then launched in again, this time pushing my foot against the cave wall behind. With an undignified sprawl, I landed on the far side. I was through. I sat up and looked around. The walls of the passage were smooth and beautifully etched, with huge scallop marks. Looking ahead, I could see that the passage was getting bigger. In the distance, almost at the limit of my beam of light, I could just make out the velvety shimmer of water. I scrambled to my feet, calling back through the rift, 'We've found it, we're into new passage, we've found the river....'*

Caves have always held a fascination. For some, the darkness has beckoned them from the comfort of daylight to its wild, remote, inhospitable and unknown depths. Without this breed of explorers, we would not have intricate cave surveys, descriptions of extraordinary cave creatures, information about past environments or beautiful photographs of cave passages. Under the surface of the earth, is a storehouse of mystery and wonder.

And perhaps the most exciting thought is that only a fraction of the earth's caves have been discovered, let alone explored.

The Caves and Limestone Scenery of the North of Ireland

Karst and conservation

Beneath our Feet

9

KARST AND CONSERVATION

By now, we can be in no doubt that our limestone landscape, both above and below ground, is limited and delicate. Because of its three-dimensional character, the surface and underground environments are inextricably linked. We have seen that the karst stores a great reserve of water. We have also seen that the landscape and caves are part of our natural and archaeological heritage, and that cave life is an important but often over-looked part of our planet's biodiversity. It is clear that there is still much to learn about our karst environment.

This limestone landscape is a vibrant living one, supporting homes and employment and generating income. It has been shaped and moulded by its inhabitants for many thousands of years. Agriculture, forestry, tourism, recreation, mineral extraction and water management are among the most important forms of activity taking place. Other than some remote sections of caves, there are few places, if any, which are truly pristine and untouched. Management of this limestone heritage not only involves thinking about the present and future, it also has to address and redress the inherited problems of past human impact.

Fig. 115 A caver examines antlers found in a remote pothole in County Fermanagh. Caves are nature's museums; they contain irreplaceable, dateable records of the past.

When we think of ways in which the limestone environment is under pressure, the highly visible examples like quarrying come to mind. But it is becoming increasingly evident that even small, seemingly minor actions can, on limestone, have significant and far-reaching effects. Above ground, it is subjected to the same pressures as the rest of our countryside. Underground, it is endangered not only by human visitors but also by the consequences effects of external actions.

LIVING AND WORKING ON THE LIMESTONE
THE DEMANDS, THE THREATS, THE IMPACT

■ QUARRYING

There is a considerable demand for quarried limestone, as it is used for a wider range of purposes than any other rock. Quarrying, however, not only aesthetically scars the landscape but also has the potential to totally destroy an area of karst. Rock blasting and vibration from machinery can cause the collapse of underground caverns. For example, in an Ulster White Limestone quarry near Glenarm in County Antrim, quarry blasting exposed a 20 metre deep shaft. Potholers explored it before it was quarried away.

Fig. 116 A quarry eats through Knockninny Hill in County Fermanagh.

The Caves and Limestone Scenery of the North of Ireland

The removal of limestone pavement for use as ornamental garden stone is a major concern. This water-worn rock has traditionally been used to create rockeries, and in the process a unique feature has been removed forever from the natural landscape. And not only is the rock unique, the pavement is a special habitat, its clefts and fissures sheltering a specific range of plants.

Britain and Ireland have some of Europe's most important areas of pavement, huge extents of which have already been devastated to satisfy landscaping fashion. Limestone pavement is now protected by laws which prohibit damage and disturbance; however, while there is demand for the stone, there is always the possibility that pavement will be extracted illegally. This threat will continue until consumers are made fully aware of the damage they are causing when they decorate their gardens with such geological masterpieces. There is no such thing as a sustainable or green source of limestone pavement; once removed, this habitat is gone forever.

The good news is that imitation water-worn rock, almost identical in appearance to the real stuff, can in fact be made simply and cheaply; the box below explains how.

Fig. 117 These pieces of limestone pavement were once part of a unique habitat.

■ HOW TO MAKE YOUR OWN LIMESTONE PAVEMENT

Instant rocks can be made by following this method thought up by gardener and conservationist Geoff Hamilton.

1. Dig a hole to act as a mould. Start with a small one until you perfect the technique, and remember you will have to be able to lift it out.

2. Line the hole with a piece of polythene. Don't worry about all the wrinkles and bends in the polythene, they will give the 'rock' a more authentic appearance.

3. Prepare the ingredients. Mix two parts coconut fibre (coir) with two parts sharp sand and one part fresh Portland cement, then add enough water to give a stiff consistency.

4. Press the mixture into the mould, working it into all the cracks.

5. Leave for a few days to set, then peel off the polythene and allow to dry.

6. You have now created a perfect piece of fake limestone for your garden without destroying original pavement.

The Caves and Limestone Scenery of the North of Ireland

■ **TOURISM, EDUCATION, AND RECREATION**

One way to appreciatie and understand limestone is to visit and see it first hand. We like to see these places for ourselves and enjoy the experience of being somewhere special or unique. But tourism puts pressure on rural areas which, by their very nature, are less able cope with huge volumes of people. We require car parks, bigger roads, places to walk, places to sit down and ways to dispose of waste. When wilderness areas are made accessible, great care has to be taken to ensure that delicate natural vegetation is not damaged or wildlife disrupted.

Karst studies are part of the curriculum for school and college students, and visits to karst landscapes are encouraged. However, guidance and careful management is essential to prevent damage to these sites from, for example, uncontrolled rock collection and sample-gathering.

The recreational use of limestone hills, cliffs and gorges for adventure activities such as mountain biking, rock climbing and abseiling also has the potential to cause wear and tear unless carried out sensitively.

All these activities, be they primarily educational or just for pure pleasure, can be carried out with maximum benefit and enjoyment for the participants and minimum impact on the natural environment. The key is responsible usage and management which ensures that the quality of the resource is sustained.

Fig. 118 Rock climbing on a steep limestone cliff. The cracks and joints in the rock along with its hardness make it a rock climber's delight.

■ **AGRICULTURE**

Most of the rural population is dependent upon agriculture. Farmers are the traditional custodians of the land. They have maintained it and shaped the limestone countryside. They have had to devise ways to carry and store water for their livestock. Their farming regimes have nurtured the flora for which limestone is renowned.

However, measures to improve farm efficiency and increase productivity may involve irreversibly altering the karst landscape. Such measures include: clearing natural vegetation, filling in dolines, the creation of tracks and bigger fields, heavy applications of slurry, re-seeding, ploughing, intensive grazing and silage production in place of hay making. Nitrate fertilisers and herbicides applied to fields travel in percolation water down through the limestone, moving particularly rapidly through thin-soiled areas and after heavy rain. In addition, it is obvious that agricultural waste and noxious chemicals, such as the effluent from silage pits and sheep dip, have to be disposed of with care to ensure that they do not end up in water courses, above or below ground.

Many of these activities alter the physical appearance of the karst landscape, and as we saw in Chapter 6, may reduce its diversity of natural flora and fauna. They may also have far-reaching effects underground in ways that might never have been imagined: pollution of ground water, the alteration of caves and cave drainage and the disruption of cave life.

FORESTRY

Fast growing coniferous plantations smother the contours of the land with a dense cloak of evergreen. In the process, this intensive planting can alter or obscure surface karst features. Today, planting is carried out sympathetically: native, deciduous trees are incorporated into the schemes and there is an awareness that the special features in the landscape, both natural and archaeological, should be left open and unplanted.

Drainage associated with forestry planting and felling can cause changes to water flow and possibly alter the pattern of underground drainage. Water chemistry is also affected, leading, for example, to an increase in acidity. This can in turn can adversely effect the rate and character of calcite deposition and can, in some cases, cause an onset of speleothem decay.

PEAT EXTRACTION

Large scale mechanical peat extraction also has a significant effect on the drainage within a karst catchment. It alters the way water runs off the land, concentrating it into certain channels. This increases the speed of drainage and also increases the amount of sediment carried along. This may lead to substantial changes within the cave systems: forcing more water in one particular direction, filling passages with peaty sediment and critically altering cave habitats.

Fig. 119 Large scale peat extraction on Cuilcagh. The peaty sludge formed as a result is channelled into the caves.

CONSTRUCTION ON LIMESTONE

Limestone may be a strong foundation on which to construct buildings and roads but it may not always be a solid one. As well as the disturbance from vibration and heavy machinery, there is the added consideration that concrete waste and other liquids are channelled underground. Run-off drainage from, for example, a newly built car park, roof or road may direct increased quantities of water into fissures in the limestone or into a particular sinkhole. These activities may have an impact on the underground environment; they may alter the pattern of underground drainage, introduce waste and may even cause localised collapse.

WASTE DISPOSAL: OUT OF SIGHT, OUT OF MIND

One of the most widespread and noticeable external threats to our karst and caves is dumping. Major cave systems and a multitude of potholes have, traditionally, served as dumps for household, commercial and farm waste. In the past, there was no refuse collection in remote rural areas. These places of beauty and seclusion were reduced to land-fill sites, choked with rubbish of all kinds: cars, vans, fridges, cookers and animal carcasses.

Fig. 120 Pothole filled to the brim with rubbish.

A survey of caves and potholes in Fermanagh in the late 1990s estimated that 25% of all sites had been used as dumps. And the refuse does not always remain where it has been dumped. Streams flowing through rubbish often carry it far underground. Black plastic, polystyrene, disposable nappies, drinks cans and bottles are found deep in caves. Inevitably, the rubbish becomes incorporated in the cave sediments.

Disused limestone quarries have also served as handy dump sites. But water filters through the rubbish and instantly flows down through the fissured rock. This is a major factor which has to be taken into consideration when land-fill sites are being planned in limestone areas.

Rubbish-filled potholes are obvious signs of pollution. However, it is the less visible contamination that is even more alarming. Eventually, polluted water emerges downhill, not as the pure water that spring wells are normally assumed to be, but as a contaminated soup. People living in limestone areas may have been blissfully unaware that the underground system which they believed to be giving them their clean drinking water was, in fact, passing on someone else's waste.

All sorts of liquids can sink into the limestone from all sorts of places. Oil and diesel may leak from storage tanks. Poorly maintained septic tanks and broken sewer pipes send liquid waste down though the limestone. Often a long time elapses before the leak is detected, in which time major damage may have been done to the store of underground waters. The discharge of such liquid wastes is, of course, against the law but because of the nature of underground drainage, it is sometimes difficult to pinpoint the source of contamination. Sewage can be a direct health hazard and is associated with the transmission of illnesses.

Fig. 121 Flowing water carries rubbish deep underground.

Drainage from areas where there may be sources of food, such as farms, stables or high density animal husbandry units may attract rats. In turn, an organism passed in rat's urine causes an unpleasant and sometimes life threatening disease in humans called Weil's Disease. The leptospira organism does not survive in dry conditions but is passed on in areas of slowly draining water, like caves.

It is not just the major streams and rivers which carry the water through the underground, but also the tiny trickles which percolate through the micro-cracks. Polluted water, filtering through to the tiny life forms sheltering in these inaccessible places, may result in the extinction of the entire fauna or cause a complete change in the types of species. There has been very little research on the degree to which polluted water changes cave life.

Limestone's thin soils give little chance for filtration. Water disappears underground rapidly, leaving no time for the evaporation of volatile compounds such as those released by petrol or diesel. The hollow nature of limestone means that water flows through rapidly, leaving little opportunity for bacteria and bugs to set about maintaining or cleaning it as they do in other environments.

We are left with the unwelcome conclusion that the likelihood of spreading contaminated water through karst is high and potentially serious.

The Caves and Limestone Scenery of the North of Ireland

THE IMPACT OF HUMANS ON THE UNDERGROUND

The underground is often altered, not only indirectly, by the activities going on above ground but also directly when we venture in to its hidden world. Unfortunately, it is sometimes all too easy for us to be unaware of our impact on these special places.

■ SHOW CAVES

Show caves provide an opportunity for us to enjoy and learn about the underground environment. The educational potential of show caves is considerable. However, show caves vary greatly in the way they are developed. In the best examples, the educational potential is maximised. The beneficial effects of this cannot be overestimated. If we can visit a cave and see for ourselves its fascinations and learn how water travels though the rock, we can gain an appreciation of why it should be conserved and how actions above ground can have such radical consequences.

The environmental impact of radically altering a cave to make it safe for public access has to be weighed against the educational and economic gains. Development may cause significant and irreversible damage to the cave unless undertaken sensitively. Inevitably, the cave will have to undergo major engineering: paths and steps may have to be put in, water courses controlled, lights installed, facilities above ground built and consideration given to the disposal of sewage so that there is no chance of leakage.

Fig. 122 The introduction of light and heat into formerly dark, cool places means that plant spores which are carried in with the human visitors will, without careful control, grow on the cave walls and calcite around the lamps, bringing green vegetation to a place where there should be none. Lamp flora not only looks unsightly but also damages calcite.

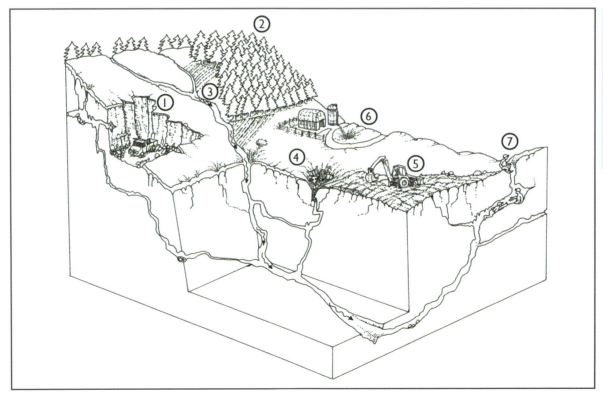

KEY

1. quarrying
2. afforestation
3. drainage
4. dumping
5. limestone extraction
6 farmyard runoff
7. caving.

Fig. 123 All sorts of human activities have an impact on the limestone above and below ground

The Caves and Limestone Scenery of the North of Ireland

The presence of high numbers of people moving and breathing, also alters the atmosphere within the cave. The temperature and carbon dioxide levels are changed and this in turn may affect stalactite and calcite growth.

The extent of tourist impact within the caves is the subject of on-going academic research. Information is constantly being gathered from studies carried out in show caves throughout the world.

Fig. 124 The Marble Arch Show Cave, County Fermanagh. Here, the use of high and moderate energy sections of the cave, coupled with relatively small, guided groups, keeps the impact from the fifty thousand visitors a year, to a minimum.

■ CAVE ENERGY LEVELS

Caves can be classified according to their energy levels: high-energy, moderate-energy and low-energy. These classifications assist decision makers, as they define which caves, or sections of caves, can withstand human usage and impact. High-energy caves regularly experience high-energy events such as flooding and the shifting of sediment banks (Fig.125a).

Moderate-energy caves are affected by nothing more dramatic than the flow of a small stream or the breath of a gentle draught, forces significantly lower in intensity to those of the high-energy floodwater (Fig.125b).

Low-energy caves only experience tiny forces such as the drip of water from the tip of a stalactite (Fig.125c).

The ability of the cave to resist the disturbances by visitors directly, relates to these energy levels. High-energy caves are robust and will remain largely unaffected by visiting humans. Moderate-energy caves are more susceptible to lasting damage by unaware visitors. Just by moving through an area, visitors are releasing similar amounts of energy to those which the cave experiences naturally. Footprints in mud will remain for hundreds of years. Low-energy sections of a cave are highly susceptible to permanent damage if entered by humans. A visitor could for example release more energy in the cave than it has experienced in a thousand years.

Fig. 125a

Fig. 125b

Fig. 125c

■ WILD CAVES

Caves which have not been developed for public access are often termed 'wild'. They are visited, almost exclusively, by cavers. Cavers are the only people with the knowledge and capability to explore, study and monitor the underground safely. However, as they are the only visitors, they may also present a potential threat to the caves.

It is a bitter pill for avid cavers to swallow to realise that, by their very presence, they are altering the places they love. As soon as a cave is entered for the first time, changes take place. The consequences will be greater in some parts of a system than others. Wading in the flowing water of a canyon passage has little effect, whereas crawling on sediment among pristine stalagmites has the potential for major disruption. A comprehensive understanding of caves helps the visiting cavers reduce their impact, so that they pass through the cave leaving nothing but the occasional knee mark or foot print. Simple rules, such as all waste being taken out of the cave, are part of the essential education for budding speleologists. Even crumbs of chocolate from an energy-giving snack bar can upset the balance and distribution of cave communities. An appreciation of the forces which shape and decorate caves is fundamental to cave exploration so that there is an awareness of the consequences of all actions.

Cavers searching for new passages also face hard decisions. They must constantly balance advancement of cave knowledge against conservation of the cave. The digging-out of sediment from a sand-filled passage might open the way into kilometres of unexplored cave. However, the flow of air through the cave may be altered, or the disturbance of the sediment layers may mean a loss of information which would help in building up an understanding of the past. Which is more important? The decision must be carefully considered.

People also experience wild caves as members of trips led by outdoor education centres and commercially-run organisations. It is obvious that the more people who enter a particular cave, the greater the impact. Difficult decisions have to be taken by the outdoor centre managers and cave leaders who plan group caving, to balance safety, cave conservation and educational needs. Regular visits by outdoor pursuits groups can cause general wear and tear on the cave. If low energy caves are visited, the accidental or intentional breakage and damage to stalactites and stalagmites is inevitable over time. The choice of high energy caves, coupled with good, safe leadership, keeps damage to a minimum.

The Caves and Limestone Scenery of the North of Ireland

■ SPELEOLOGICAL UNION OF IRELAND

Ireland has its own national, co-ordinating body for caving, the Speleological Union of Ireland (SUI), which seeks to educate and inform those entering caves. It has adopted a conservation and access policy and a cave leader training scheme which has "Safety of the Cave" as a major section of the syllabus.
www.cavingireland.org

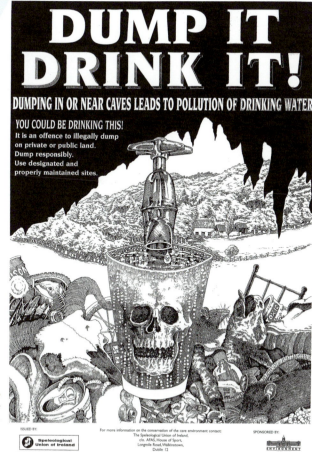

Fig. 126 A cave conservation poster produced by SUI. An example of the educational efforts of SUI to educate the non-caving public.

■ WHO OWNS CAVES?

A question which is often asked is, "Can anyone say that they own a cave?" This is a complicated query when the many issues involved are considered: for example,

- can an empty space be owned by anyone?
- how deep into the earth's crust can anyone lay claim?
- is there any economic value related to the cave?
- can caves be owned even if they haven't been discovered?

The Law currently regards caves as part of the overlying land so that the owner of the land above owns the land beneath. But this can become very complicated; if the same cave extends under adjacent land with a different owner, then they in turn own the cave underneath, even if there is in no entrance to it on their land. So one large cave system can theoretically have many owners - a management nightmare!

■ STATUTORY CONSERVATION MEASURES

The great importance of these karst and cave systems, together with their associated habitats and landscapes, is reflected in the range of conservation designations that have been made. Underlying all of these designations, is the desire by government to work closely with landowners and other parties with an interest in these sites, to achieve their long-term sustainable management.

The need for conservation designations generally, has come about through the widespread recognition that human impact on the countryside has led to the severe reduction in or the entire loss of certain species, habitats and other features of conservation importance.

A brief explanation of each relevant designation follows.

Areas of Special Scientific Interest (ASSI) are sites that have been scientifically surveyed and are deemed to be of great conservation importance. By and large, they remain in private ownership. The part played by landowners in maintaining these sites is fully recognised and appreciated. Notification of an ASSI is made under national legislation – the Nature Conservation and Amenity Lands (Northern Ireland) Order 1985 (as amended).

The Special Area of Conservation (SAC) designation reflects the importance of certain habitats in a European context. Such sites are designated as ASSIs prior to the SAC notification; the additional designation reflects the international significance of the site. Selection of SACs are a response to the European Habitats Directive and is implemented in Northern Ireland through The Conservation (Natural Habitats etc.) Regulations (Northern Ireland) 1995.

Nature Reserves (NR or NNR) are sites of importance for their conservation features, which are managed for conservation and are generally in the ownership of organisations who are devoted to conservation. They provide special opportunities for study and research, as well as for recreation. Such sites are also designated under the Nature Conservation and Amenity Lands (Northern Ireland) Order 1985.

Farmers with land within Environmentally Sensitive Areas (ESA) can receive payments for managing their land in a less intensive manner, thereby benefiting the native plants and animals found there. ESAs are designated and administered by the Department of Agriculture and Rural Development.

For its small area, Northern Ireland has a great variety of scenic countryside. Large areas of landscape of distinctive character and special scenic value have been designated Areas of Outstanding Natural Beauty (AONB). This designation is designed to protect and enhance the qualities of each area, and to promote their enjoyment by the public. Once again, the Nature Conservation and Amenity Lands (Northern Ireland) Order 1985 is the relevant piece of legislation.

See the following web pages for more information on site designations
www.ehsni.gov.uk/NaturalHeritage/StaticContent/sitedesignation.htm

The Caves and Limestone Scenery of the North of Ireland

■ *MANAGING A SPECIAL ENVIRONMENT*

How can this unique karst environment remain both a vibrant, living landscape and one that is treasured and protected for the future? This is a challenging management problem, when the many threats and demands are considered.

The importance of karst and caves is well recognised world-wide and general guidelines for cave and karst protection have been written by the International Union for Conservation of Nature and Natural Resources. But what about closer to home? Everyone who lives or works on or near limestone, or who visits it, will consciously or unconsciously have some impact on it, but the people who have direct influence on our limestone heritage are the people who own and work the land, the statutory organisations, and the cavers. The statutory organisations include such local and national government departments as the Environment and Heritage Service. These various bodies have influence on the karst countryside: the Planners on development on karst, the Water Service on the control of water courses and pollution, the Department of Agriculture on farmland management and Local and District Councils on tourism, recreation and waste disposal.

These people must work together to weigh up the conflicting demands, evaluate the effects and come up with workable solutions which protect both livelihoods and landscape.

The Environment and Heritage Service (DOE) has carried out an Earth Science Conservation Review which included a detailed description of Northern Ireland's significant karst and caves, which was completed in 1995. This stock-take acts as a base-line for the on-going monitoring of its quality and diversity. Sites which are recognised as being particularly special, representative or vulnerable may be granted further protection by law.

Conservation designation demonstrates that a landscape, or feature within a landscape, such as a natural rock arch, an area of limestone pavement, or perhaps a particularly fine wildflower meadow, is worthy of recognition and special protection. Similarly underground, a cave with superb stalactites or banks of undisturbed sediments may be designated. Ideally, to ensure complete protection, the whole surface catchment area of a cave, and not just the cave passage alone, should, if possible, be included in the designation.

But, of course, it is the management of the site which assures its preservation. All the designations in the world cannot assure protection if everyone involved is not committed to its conservation.

To look at how the process of sustainable management works, we can take the example of a cave with superb stalactites. When it has been ascertained that the stalactites are particularly special, the cave may be given Area of Special Scientific Interest (ASSI) status. Once designated, the landowner of the land above the cave system cannot change the cave or the land above the cave without permission. It is possible that some changes the owner may wish to make would have no impact on the cave; however, it is equally possible that a change, for example, to drainage, could starve the stalactites of the water supply essential for their growth and well-being. The Environment and Heritage Service decides if any proposed changes have the potential to adversely affect the scientific importance of the site. Cavers, as the only visitors to the stalactites, have the responsibility of looking after them and monitoring their condition. They cannot, however, enter the cave without the landowner's agreement. And so an interconnecting network of communication and responsibility is created between owner, legislator and caver. This web of law, responsibility, trust and permission is a complex one and unfortunately easily broken. It is not difficult to imagine how the action of one group or person within the network can cause a breakdown; it is the stalactites which may suffer as a result.

The management of the web relies first on the groups involved understanding the importance of the caves and karst of the region and second, on open discussion between all of them to develop a sustainable management plan. If the opinion of one of the groups is ignored, the result will inevitably mean conflict and a break in the web. It is also inevitable that, to achieve a sustainable management plan, compromises will have to be made by all. Any plan needs to be continually monitored, assessed and modified by the whole management team.

We are fortunate that limestone forms part of our natural heritage. The protection of this special landscape can only be assured if we have an appreciation of its importance, an understanding of the ways in which it can be damaged, and a mutual desire to manage it sustainably, so that it can be enjoyed by future generations.

The Caves and Limestone Scenery of the North of Ireland

The Caves and Limestone Scenery of the North of Ireland

WHERE TO EXPLORE LIMESTONE

■ MAPS AND GUIDES

O.S. (N.I.) Discovery Series 1:50,000
Co. Antrim: Sheets 5 & 9
Co. Fermanagh: Sheets 17 & 26

The **Geological Survey of Northern Ireland** and the **Geological Survey of Ireland** have produced the following guides in their Landscapes from Stone series:

Explore Cuilcagh
Explore Upper Erne
Explore West Breifne
Walk Cuilcagh
A Guide to the Scenic Landscapes and Rocks of Ireland (North).
A Story through Time - The Formation of the Scenic Landscapes of Ireland (North)

Contact the Geological Survey of Northern Ireland, 20 College Gardens, Belfast BT9 6BS tel: 028 9066 6595 for further details or view their web pages on
www.bgs.ac.uk/gsni/GSNIirish_geology.htm or www.gsi.ie/active/tourism.htm

Some of the places to see Antrim's limestone are described in 'A Geological Excursion Guide to the Causeway Coast' by Paul Lyle, one of the books in the Environment and Heritage Service's Interpretation Series.

Contact Environment and Heritage Service, 35 Castle Street, Belfast BT1 1GU tel: 028 9025 1477 for further details or view their web pages on
www.ehsni.gov.uk/NaturalHeritage/

■ CO. FERMANAGH

The Marlbank Scenic Loop Drive
Sign-posted off the Florencecourt-Blacklion road. The road leads uphill through the limestone landscape and passes within sight of many features of karst interest. Views to Cuilcagh Mountain, Belmore Mountain and the Fermanagh lowlands. The nature reserves listed below lie on the loop, or along the Florencecourt-Blacklion road which runs along the base of the limestone, connecting each end of the loop.

Killykeeghan and Crossmurin National Nature Reserve, Marlbank.
Marlbank scenic loop road.
Car park. O.S. (N.I.) F 108 341
Way marked trail through typical karst scenery and sites of archaeological interest.

The Caves and Limestone Scenery of the North of Ireland

Interpretative centre
Open May to September

See web page for further site information
www.ehsni.gov.uk/NaturalHeritage/StaticContent/crossmurrinnr.htm

Marble Arch National Nature Reserve, Cladagh Glen
Florencecourt-Blacklion road
Car Park at Cladagh bridge. O.S. (N.I.) F 125 357
Open all year round
Footpath follows the Cladagh River upstream through moist ash woodland, passing the Cascades River rising and other sites of karst interest on the way. It leads over the limestone arch, the natural rock bridge which gives the glen its name, and on to the Marble Arch Show Cave Visitor Centre.

See web pages for further site information -
www.ehsni.gov.uk/NaturalHeritage/StaticContent/marblearchnr.htm

Hanging Rock and Rossaa National Nature Reserve
Florencecourt-Blacklion road
Roadside car park. O.S. (N.I.) F 106 366
Moist ash woodland at foot of limestone cliff.
Short trail leads to Hanging Rock River rising
Open all year round

See web page for further site information
www.ehsni.gov.uk/NaturalHeritage/StaticContent/hangingrocknr.htm

Cuilcagh Mountain Park
Marlbank scenic loop road
Car park O.S. (N.I.) F 121 335
Track and way-marked hiker's trail to summit of Cuilcagh Mountain. The park includes limestone scenery and extensive expanses of blanket bog.
Interpretative centre based at Marble Arch Show Cave, open seven days a week from Easter to September.

Florence Court Forest Park
Florencecourt road
Car park O.S. (N.I.) F 179 348
Way-marked hiker's trail leads from the forest park, through the limestone of East Cuilcagh, and across the extensive bogland to the summit of Cuilcagh Mountain.

The Caves and Limestone Scenery of the North of Ireland

Gortalughany Scenic Drive
Sign-posted off Enniskillen -Swanlinbar road, 2 miles from Swanlinbar.
Car park O.S. (N.I.) F 168 301
View-point with panoramic vistas to Cuilcagh Mountain, Benaughlin Mountain, Knockninny Mountain and the Lough Erne basin.

CO. ANTRIM

Linford, Carncastle
Car park on the Ulster Way O.S. (N.I.) D 333 073
Small area of Ulster White Limestone. From the road, west of the car park, karst detectives with a keen eye may be able to make out a dry valley, sinks and dolines. Ancient enclosures are clearly visible.

Belshaw's Quarry, Lisburn
Sign-posted off the Lisburn to Antrim road
Car park. O.S. (N.I.) D 229 671
Open all year round
Trail leads through disused Ulster White Limestone quarry. Basalt intrusions clearly visible in cliff faces.

See web page for further site information
www.ehsni.gov.uk/NaturalHeritage/StaticContent/belshaw.htm

Loughaveema-The Vanishing Lake
Cushendun - Ballycastle road
O.S. (N.I.) D 205 360

UNDERGROUND IRELAND

If you want to experience caves at first hand then the best way to do so is by visiting a show cave. And if, after having the show cave experience, you want to delve deeper into 'wild' caves, you should never consider heading off on your own. Cave exploration needs training, experience and the proper equipment. The best way to enjoy caving and to ensure the conservation of the cave environment, is to go to an outdoor education centre which offers caving trips led by leaders holding the national caving qualifications.

SHOW CAVES IN IRELAND

Marble Arch Caves, Marlbank, Co. Fermanagh
Set in the midst of the Fermanagh limestone landscape, this tourist cave has an interpretative centre explaining the cave and the surrounding landscape. The cave tour, which last an hour and a half, retraces the footsteps of its first explorer, Martel, and

includes a short boat trip. It is advisable to phone and book before visiting as the cave may close for safety reasons following heavy rain.
Open from May to September.

See web page for further site information
www.fermanagh.gov.uk/tourism/attractions/attractions-main.htm

Aillwee Cave, Ballyvaughan, Co. Clare
Deep in the heart of the unique Burren landscape, this cave boasts brown bear hibernation pits close to the entrance. It has a good example of a half-tube in the roof, showing the earliest stages in the formation of a cave.

Dunmore Cave, Co. Kilkenny
Located just north of the city of Kilkenny, this tourist cave run by the Office of Public Works, is of historical as well as scientific interest. It is mentioned in the Annals of the Four Masters, describing the slaughter near the cave of a thousand local people by the Vikings in AD928. This was confirmed during the development of the cave for tourism, when a hoard of coins were found dated to that time. It has impressive chambers and a huge stalagmite column.

Crag Cave, Castleisland, Co. Kerry
Discovered by cavers in the early 1980s, the story of its exploration is an exciting one; a cave diver swam through a water-filled passage to find extensive open passages lying beyond. Shortly after discovery, it was developed as a show cave. It is particularly noted for its array of delicate straw stalactites.

Mitchelstown, Co. Tipperary
This cave near Ballyporeen, was discovered in 1833 during quarrying. The cave is a maze of passages typical of one formed principally below the water table. The tourist tour takes visitors through a series of impressive chambers linked by short passages richly decorated with stalactites, stalagmites and columns.

SOME FURTHER READING

Burns, G., Fogg, T., Jones, G Ll., & Kelly, J. G. (1997). The Caves of Fermanagh and Cavan. Lough Nilly Press, Fermanagh.

Chapman, P. (1993). Caves and Cave Life. The New Naturalist Library. Harper Collins, London.

Coleman, J.C. (1965). The Caves of Ireland. Anvil Books, Tralee.

Department of the Environment for Northern Ireland, (1991). Fermanagh Its Special Landscapes. HMSO.

Farr, M. (1984). The Great Caving Adventure. The Oxford Illustrated Press.

Fogg, T. & Kelly, J.G. (1995). Karst Geomorphology of Northern Ireland. Environment and Heritage Service Earth Science Conservation Review. Unpublished report to Environment and Heritage Service, DOE.

Ford, D., & Williams, P. (1989). Karst Geomorphology and Hydrology. Chapman and Hall, London.

Geological Survey of Northern Ireland (1991). 1:50,000 series. Derrygonnelly. Sheet 44, 56 and 43. Solid edition. British Geological Survey.

Geological Survey of Northern Ireland (1998). Geology of the country around Derrygonnelly and Marble Arch. British Geological Survey.

Geological Survey of Ireland (1962). 1:750,000 Geological map of Ireland. 3rd edition, 1985. Government of Ireland.

Geological Survey of Ireland (1996). Geology of Sligo-Leitrim. Department of Transport, Energy and Communications.

Hamond, F. (1991). Antrim Coast and Glens Industrial Heritage. Department of the Environment for Northern Ireland. HMSO. Belfast.

IUCN World Commission on Protected Areas (1997). Guidelines for Cave and Karst Protection.

Jackson, D.D. (1982). Underground Worlds. Planet Earth. Times-Life Books, Amsterdam.

Jennings, J.N. (1985). Karst Geomorphology. Blackwell, Oxford.

Lyle, P. (1996). A Geological Excursion Guide to the Causeway Coast. Environment and Heritage Service.

Keenan, P.S. (1995). Written in Stone. Geological Survey of Ireland.

Lowe, D. & Waltham, A.C. (1995). A Dictionary of Karst and Caves. Cave Studies Series Number 6, British Cave Research Association.

Lyon, B. (1983). Venturing Underground. EP Publishing, West Yorkshire.

Mallory, J.P. & McNeill, T.E. (1991). The Archaeology of Ulster. Institute of Irish Studies, The Queen's University of Belfast.

McKeever, P.J. (1999). A Story Through Time. Geological Survey of Ireland

National Caving Association, (1997). Cave Conservation Handbook.

Parker, N and Keaveney, E. (2000). 25 walks in Fermanagh. Fermanagh District Council.

Wilson, H.E. (1972). Regional Geology of Northern Ireland. HMSO. Belfast.

GLOSSARY OF KARST AND CAVE TERMS

All terms in bold are also in the glossary.

Acid rain: rain water or snow which has become slightly acid because it has dissolved sulphur dioxide gas held in the atmosphere. Sulphur dioxide originates mainly from oil and coal fired power stations (see **carbonic acid**).

Active: description of a cave passage or cave with running water in it. Fig. 125a.

Aggressive water: water containing an active chemical ingredient, such as **carbon dioxide**, which enables it to corrode limestone rock. Fig. 9.

Anastomoses: intricately branched and braided system of tubelets formed by solution along planes of weakness in **phreatic** condition in the early stages of cave formation. Fig. 26.

Aquifer: layer of rock which holds water, allows water to move through it and is the source of water for wells and springs. Fig. 13.

Aven: hole in the roof of a cave passage that may either close into impenetrable fissures but still take water or may be an open route into the cave. Seen from above it would be called a **shaft**. Figs. 68 and 76.

Bed: layer of sedimentary rock lying between two **bedding planes**. Fig. 7.

Bedding plane: line of weakness separating two beds of limestone (and other sedimentary rocks). Bedding planes often provide a pathway for water movement through limestone and may be opened out to form bedding plane caves. Figs. 7 and 31.

Bio-speleology: scientific study of life within caves.

Blind valley: valley that stops abruptly at a point where its stream sinks, or once sank, underground. Fig. 21.

Botryoidal calcite: globular **calcium carbonate speleothems** also called popcorn or cave coral.

Breakdown: pile of rock filling all or part of a cave passage after the collapse of part of the walls or ceiling. Fig. 41.

Bridge: natural bridge or arch often formed at the entrance of a cave by partial collapse leaving an isolated roof section. Examples are the Marble Arch and the Greenan Arch. Fig. 70.

Calcareous: description of rock that consists of **calcium carbonate** or lime.

Calcite: mineral formed of calcium carbonate ($CaCO_3$) which is the main constituent of limestone and of most **speleothems** such as **stalactites** and **stalagmites**.

Calcium carbonate: compound which occurs naturally as the minerals calcite and aragonite. It is the major component of **carbonate rocks** and is also common in other sedimentary rocks. Its chemical formula is $CaCO_3$.

Canopy: flowstone which projects from a cave wall. It may result from the removal of underlying sediments or from the steady outward development of flowstone creating an overhang. Fig. 66.

Canyon passage: often referred to as a **vadose** canyon formed by the down cutting of a stream.

Carbide: short for calcium carbide (Ca_2C). A chemical which gives off acetylene gas when mixed with water. The fuel for carbide lamps.

Carbon dioxide: naturally occurring gas which when dissolved in water produces **carbonic acid**.

Carbonate rocks: rocks like **limestone** or **marble** which consist mostly of one or more carbonate minerals i.e. **calcite** or aragonite.

Carbonic acid: weak natural acid formed by the absorption of **carbon dioxide** in soil and rain water. The dissolving of **calcium carbonate** by this acid is the predominant reaction in the **karst** process. Fig. 9.

Carboniferous: period of geological time from approximately 360 to 290 million years ago. The Fermanagh limestones are of Carboniferous age. Fig. 46.

Catchment: area from which a stream or cave receives or received its water.

Cave: a natural underground opening in rock large enough for human entry.

Cave formations: see **speleothem**.

Cave pearls: round **calcite** pebbles which develop by deposition around a central grain. They form where water drips and constantly turns the stone allowing even deposition. Fig. 37.

Cave system: series of inter-linked cave passages.

The Caves and Limestone Scenery of the North of Ireland

Cavernicole: species which lives in a cave habitat and can complete its life cycle there.

Chalk: soft white or greyish type of porous limestone.

Chamber: enlarged area of cave passage.

Chemical erosion: see **dissolution**.

Chert: rock composed of silica (SiO_2) which commonly occurs in nodules or bands in limestone. Similar to flint. Fig. 79.

Clints: single block in **limestone pavement**. Fig. 19.

Closed depression: see **doline**.

Coccolith: calcareous skeletal plate of a microscopic marine organism. A major component of **chalk**. Fig. 47.

Column: **stalactite** and **stalagmite** that have grown until they join roof to floor. Fig. 33.

Corrasion: mechanical erosion of rock surfaces by rock debris carried in water.

Corrosion: chemical erosion of rock by solution.

Covered karst: karst landscape where much of the development is beneath soil cover in contrast to bare karst where features develop on uncovered bedrock. Fermanagh has predominately covered karst while the Burren, Co. Clare has mostly bare karst. Fig. 17.

Curtain: sheet of calcite hanging from a cave roof or wall formed when water runs down an inclined roof before dropping free. Often banded with mineral impurities. Fig. 36.

Dip: angle between the horizontal and an inclined bedding plane.

Dissolution: natural process of dissolving a solid. In the karst process, the dissolving of carbonate rock to create a solution of calcium and bicarbonate ions in water. Fig. 9.

Doline: conical-shaped depression in the karst land surface with no surface downhill drainage. Also called closed depressions. Figs. 15 and 16.

Dolomite: mineral composed of $CaMg(CO_3)_2$ that has similar properties to **calcite**. Limestone with a significant proportion of dolomite is called dolomitic limestone.

Draught: used to describe the natural movement of air through a cave as a result of differences in barometric pressure or temperature at two entrances, or different areas of a cave system.

Dripstone: general term for calcite deposits resulting from precipitation from water.

Dry valley: valley cut by a river or stream, then abandoned and left dry when underground drainage developed. Fig. 20.

Dye tracing: see **Water tracing**.

Dyke: vertical or sub-vertical wall of igneous rock injected into a fracture or fault in the earth's crust. Fig. 50.

Eccentric: see **helictite**.

Elevation: term used for a type of cave survey which takes a vertical cut or sideways view of the cave (see **plan**). Figs. 77 and 84.

False floor: sheet of flowstone left when the sediment on which it formed has been washed out from beneath. It may be a complete bridge or just projecting ledges from the cave wall.

Fault: fracture where the two sides of the fracture are displaced relative to each other. Figs. 7 and 77.

Fissure: any break within a rock mass. In karst rocks it refers to **bedding planes**, **joints** and **faults** which may be open, or could be opened by dissolution, to provide a route for water movement. Fig 7.

Flint: form of silica, similar to **chert**, occurring in **chalk**. Fig. 45.

Flood pulse: sudden peak stream flow that occurs after the onset of heavy rain or snow thaw.

Flowstone: **calcite** deposited on walls or floors of caves. Fig. 39.

Fluroscein: harmless dye which when dissolved in water is bright green. Commonly used to trace underground water. Fig. 12.

Fossil: remains, imprint or trace of an animal, plant or microbe which has been buried and preserved in sedimentary rocks. Figs. 4, 56 and 57.

Fossil karst: **karst** landforms formed in past eras and preserved within the rock sequence. Also known as **paleokarst**. Figs. 24 and 82.

Geomorphology: study of the physical features of the earth or the arrangement and form of the earth's crust and of the relationship between these physical features and the geological structures beneath.

Glaciokarst: **karst** landscape that was glaciated and displays land forms of glacial origin. Figs. 69 and 70.

Gour: barrier of flowstone which dams a pool known as a gour pool. Their development is self-perpetuating as the thin film of water overflowing the dam precipitates more **calcite**. Fig. 38.

Groundwater: general term used to refer to any underground water but strictly refers to water in the **phreatic zone**. Figs. 11 and 13.

Gryke: solution widened **fissures** in **limestone pavement** which surround a **clint**. Also spelt grike. Figs. 18 and 19.

Half tube: semi-circular channel in a cave roof. Half tubes are what remains of a complete phreatic tube after the bottom half has been removed by cave enlargement or collapse. Half tubes are evidence of a cave's early **phreatic** development. Fig. 26.

Helictite: usually small twisted and contorted calcite growths which defy gravity. Distortion may be the result of abnormal crystal growth, impurities or draught - guided water drops. Figs. 35 and 75.

Hydrology: study of water flow above and below ground.

Inception: earliest stage of cave formation when water first starts to move through and enlarge fissures in carbonate rocks. Figs. 25 and 26.

Joint: fracture where no displacement has occurred. Fig. 7.

Karren: German term used to describe small-scale limestone surface features. Similar to French lapiaz. Fig. 19.

Karst: general term which describes scenery and land forms developed on a soluble rock with underground drainage through caves.

Key-hole passage: classic cave cross-section shape consisting of a **phreatic** tube with a **vadose** canyon cut into its floor. It indicates two clear phases in the development of the passage. Fig. 30.

Laminar flow: type of water-flow through fine fissures in which no eddying or mixing occurs (see **turbulent flow**). Dominant in the **inception** stage of cave development. Fig. 25.

Lamp flora: alien plant growth associated with artificial light and heat produced by lighting systems in show caves. Fig. 122.

Limestone: sedimentary rock containing at least 50% **calcium carbonate**.

Limestone pavement: bare limestone surface fretted by **karren**. The exposure of the limestone is a glacial process and limestone pavements are **glaciokarst** features. Figs.17, 18 and 19.

Macrocaverns: **bio-speleological** term for the larger cave habitats, occupying the size range 200mm or more.

Marble: metamorphosed and re-crystallised **carbonate rock**. Caves can develop in marble.

Maze cave: cave system consisting of a generally horizontal network of inter-connecting passages which have developed over the same time period. Boho Caves, Co. Fermanagh are a classic example of a joint-guided maze cave. Fig. 79.

Mesocaverns: **bio-speleological** term for cave habitats ranging in size from 1 -200mm.

Microcaverns: **bio-speleological** term for smallest cave habitats, less than 1mm in size.

Moonmilk: lumpy white masses found on cave walls consisting of carbonate minerals, sometimes in association with bacteria and algae. Pasty when wet and powdery when dry.

Mud mound: accumulations of lime-muds sometime attributed to algal colonies. The resulting fine-grained limestones form the knolls which characterise the Fermanagh limestone scenery. Figs. 43 and 44.

Palaeokarst: see **fossil karst**. Figs. 24 and 82.

Percolation water: water moving slowly through the micro-fissures in limestone.Fig. 9.

Permeability: ability of rock to allow the passage of water

Photokarren: light orientated **phytokarst**, usually in the form of pinnacles.

Phreas: zone of total water saturation below the **water table**. Figs. 25 and 26.

Phreatic: term applied to a passage formed in water-filled conditions. Fig. 27.

Phytokarst: small scale karst features weathered by organic acids from growing plants. In temperate climates the plant life is usually algae. Also known as biokarst.

Pitch: vertical shaft but more particularly the route taken by cavers to descend or ascend it. Figs. 68 and 76.

Plan: term used for a type of cave survey which takes a horizontal cut or bird's eye view of the cave (see **elevation**). Fig. 77.

Poll: Irish for cave.

Pothole: single shaft or entrance to a cave system. Also used to describe an entire cave system that is mostly vertical. Fig. 14.

Poul: see **poll**.

Quaternary: period of geological time from approximately 1 million years ago to the present time. Fig. 46.

Radon: naturally occurring carcinogenic gas deriving from the breakdown of radioactive materials in rock.

Relict cave: inactive sections of cave passage abandoned by the water that formed them. Often referred to, incorrectly, as fossil caves. Fig. 30.

Resurgence: term used for the point where a stream or river emerges from a cave system. Also known as spring or rising. Fig. 22.

Rift: high, narrow and generally straight cave passages developed on major **joints** or **faults**. Fig. 31.

Rimstone pool: pools surrounded by **gours**. Also known as gour pools. Fig. 38.

Rising: see **resurgence**.

Scallop marks: spoon-shaped hollows carved on cave passage walls by flowing water. The size of the marks is an indication of the speed of water flow that formed them. The smaller the marks the faster the flow. Fig. 28.

Shaft: vertical or steeply inclined cave passage. Figs. 68 and 76.

Shakehole: northern English term for a subsidence **doline** formed in glacial till overlying limestone. Fig. 16.

Shale: fine grained sedimentary rock produced from clay. Shale is often found in bands which create insoluble layers inter-bedded with limestones.

Sink: point where surface stream goes underground. Also called swallet. Fig. 14.

Slickenside: lined and polished surface on a **fault** plane where the two sections of rock moved across each other. Fig. 77.

Solution: see **dissolution**.

Solution hollows: form of **karren** consisting of cup-shaped hollows. Fig. 19.

Solution runnels: form of **karren** consisting of smooth shallow channels separated by sharp angular ridges found on bare karst. Figs. 18 and 19.

Speleogenesis: birth and development of a cave from inception to its collapse and complete removal.

Speleologist: cave scientist. In France known as a spéléologue and in the U.S. as a spelunker.

Speleology: study of caves.

Speleothem: general term for all cave mineral deposits including stalactites, flowstone etc.

Spring: point where underground water emerges on to the surface but not specific to karst. Fig.22.

Stalactite: hanging, carrot-shaped speleothem, normally calcite. Figs.33 and 34.

Stalagmite: speleothem formed by upward growth from the floor as a result of carbonate-saturated water dripping from the cave ceiling above. Fig. 33.

Straw: hollow, thin-walled tubular stalactite (like a drinking straw) which may grow to a length of several metres. Fig. 32.

Sump: section of cave passage flooded to the roof. Also called siphon.

Swallow hole: the point at which surface water disappears underground. Fig. 14.

Terraces: flat areas of land between slopes.

Tertiary: period of geological time from approximately 60 to 1.6 million years ago. Preceding the Quaternary period. Fig. 46.

Travertine: a hard version of **tufa**.

Troglobites: **cavernicole** which shows morphological features such as loss of eyes, which suggest that it has undergone a long history of cave habitation. Fig. 99.

Troglodyte: human cave dweller.

Troglophile: **cavernicole** which is known to complete its life cycle in non-cave habitats as well as cave habitats.

Trogloxenes: species which occurs in caves but does not complete its life cycle there.

Tube passage: cave passage with a circular cross-section formed when full of water, allowing equal solution all around. When no longer water-filled referred to as relict **phreatic** tubes. Fig. 27.

Tufa: soft calcareous mineral deposited by water as it flows out of limestone, occurring as a result of the loss of carbon dioxide from the calcium-saturated water. Often occurs where plants or algae remove the carbon dioxide and so promote carbonate precipitation. Normally refers to deposits found outside caves and therefore excludes **speleothems**. See also **travertine**. Fig. 94.

Turbulent flow: water flow pattern which develops due to friction with the conduit wall producing eddies. This occurs when **laminar flow** has increased the fissure size.

Turlough: karst depression that is seasonally dry or flooded depending on groundwater levels. Derived from the Irish for dry lake. Figs. 23 & 80.

Vadose: zone with air above water. Fig. 29.

Water table: the level within a rock mass below which all spaces are totally water-filled, known as the flooded or **phreatic** zone. The free-draining **vadose** zone is above. In limestone, the water table may be variable and localised. Fig. 11.

Water tracing: technique for confirming underground links between river sinks and risings by putting traceable substances likes dyes into the water at the sinks and detecting it at points downstream. Fig. 12.